Helmut Eschrig
The Particle World of Condensed Matter

EAGLE 024:
www.eagle-leipzig.de/024-eschrig.htm

Edition am Gutenbergplatz Leipzig

Gegründet am 21. Februar 2003 in Leipzig.
Im Dienste der Wissenschaft.

Hauptrichtungen dieses Verlages für Forschung, Lehre
und Anwendung sind:
Mathematik, Informatik, Naturwissenschaften,
Wirtschaftswissenschaften, Wissenschafts- und Kulturgeschichte.
Die Auswahl der Themen erfolgt in Leipzig in bewährter Weise.
Die Manuskripte werden lektoratsseitig betreut, von führenden
deutschen Anbietern professionell auf der Basis Print on Demand
produziert und weltweit vertrieben. Die Herstellung der Bücher
erfolgt innerhalb kürzester Fristen. Sie bleiben lieferbar; man kann
sie aber auch jederzeit problemlos aktualisieren.
Das Verlagsprogramm basiert auf der vertrauensvollen
Zusammenarbeit mit dem Autor.

Bände der Reihe "EAGLE-LECTURE"
erscheinen seit 2004 im Wissenschaftsverlag
"Edition am Gutenbergplatz Leipzig" (EAGLE).

Jeder Band ist inhaltlich in sich abgeschlossen.

EAGLE-LECTURE: www.eagle-leipzig.de/lecture.htm

Helmut Eschrig

The Particle World of Condensed Matter

An Introduction to the Notion of Quasi-Particle

EAG.LE Edition am Gutenbergplatz
Leipzig

Bibliografische Information der Deutschen Bibliothek
Die Deutsche Bibliothek verzeichnet diese Publikation in der Deutschen Nationalbibliografie; detaillierte bibliografische Daten sind im Internet über http://dnb.ddb.de abrufbar.

Prof. Dr. rer. nat. habil. Helmut Eschrig

Born 1942 in Thierfeld / Saxony. After a study of Engineering correspondence study of Physics at Dresden University of Technology (DUT). 1972 PhD. From 1969 to 1975 assistant at DUT, Institute of Theoretical Physics. From 1975 to 1991 scientist at the Central Institute for Solid State Physics and Materials Research, Dresden.
1991 habilitation. 1992 founding director, since 1998 Scientific Director of the Leibniz Institute for Solid State and Materials Research Dresden (IFW). Since 1992 Professor for Solid State Physics at DUT, Institute of Theoretical Physics.
External scientific member of the Max-Planck Society, member of Leopoldina and of the Saxonian Academy of Sciences Leipzig.

Erste Umschlagseite:
Fermi surface of cubic $MgCNi_3$, a new superconductor; left: hole sheets, right: electron sheets. The color code indicates the Fermi velocity, increasing from blue through green to red; yellow and gray: inner side of the sheet.

Vierte Umschlagseite:
Dieses Motiv zur BUGRA Leipzig 1914 (Weltausstellung für Buchgewerbe und Graphik) zeigt neben B. Thorvaldsens Gutenbergdenkmal auch das Leipziger Neue Rathaus sowie das Völkerschlachtdenkmal.

Für vielfältige Unterstützung sei der Teubner-Stiftung in Leipzig gedankt.

Warenbezeichnungen, Gebrauchs- und Handelsnamen usw. in diesem Buch berechtigen auch ohne spezielle Kennzeichnung nicht zu der Annahme, dass solche Namen im Sinne der Warenzeichen- und Markenschutz-Gesetzgebung als frei zu betrachten wären und von jedermann benutzt werden dürften.

EAGLE 024: www.eagle-leipzig.de/024-eschrig.htm

Das Werk einschließlich aller seiner Teile ist urheberrechtlich geschützt. Jede Verwertung außerhalb der engen Grenzen des Urheberrechtsgesetzes ist ohne Zustimmung des Verlages unzulässig und strafbar. Das gilt besonders für Vervielfältigungen, Übersetzungen, Mikroverfilmungen und die Einspeicherung und Verarbeitung in elektronischen Systemen.

© Edition am Gutenbergplatz Leipzig 2005

Printed in Germany
Umschlaggestaltung: Sittauer Mediendesign, Leipzig
Herstellung: Books on Demand GmbH, Norderstedt

ISBN 3-937219-24-2

Preface

Science in our days gets more and more specialized, and often even the actors in different fields of the same discipline do not find a common language any more. Horizons narrow with the danger of ending up in 'provincialism in the space of knowledge'. On the other hand, probably the most appropriate characterization of physics as a discipline is the aim of a unified understanding of nature. From time to time chains of events appear in physics where this unified approach shines up in a bright spotlight on the scene, and then again it drops out of the focus in the hard work of solving complex tasks.

Such a splendid period of mounting a peak and widening the horizon from day to day were the fifties and early sixties of twentieth century in quantum field theory with that fascinating cross-fertilization between Elementary Particle and Solid State Physics. On the side of Solid State Theory it layed the ground for the quasi-particle approach, so successful during the sixties and seventies. After a kind of stock taking and putting more emphasis on the limitations of this approach, which of course exist and have to be taken into consideration, in recent time new kinds of exotic quasi-particles were discovered and the whole approach has by no means lost its topicality, proving the vivid potentials of field theory.

In the middle of the eighties, Musik Kaganov and myself started a review project focusing on the underlying principles and the deeper understanding of the approach. This project was also triggered (at least as regards me) by our feeling in discussion with colleagues that quite not all of them were well aware of those underlying principles which we consider fundamental to the Theory of Solids. Then, the political changes came in Germany demanding a larger effort on my side in the reorganization of research in its eastern part, and Musik left Russia in 1994. The project remained unfinished.

In the summer term 2004, I read a three-hours one-term introductory course in Field Theory for the Solid State, and in preparing for it I came back to the matter as we left it in 1988. Although I now had to revise and finish the text without Musik, it is of course further on in large parts inspired by him. (The reader consulting publications by M. I. Kaganov will find out in which parts in particular.) Since we did not have a chance to discuss the result in detail, we agreed that I should now present the text as a single author. Particularly, I am now responsible for the overall style of presentation, and, of course, for all errors. The text would, however, never have been accomplished without the long period of our discussions and the honor and delight of a life-long friendship with Musik Kaganov.

Dresden, November 2004   Helmut Eschrig

Contents

Introduction		**9**
1	**Technical Tools**	**13**
1.1	Résumé on N-Particle Quantum States	13
	1.1.1 Schrödinger Representation	15
	1.1.2 Spin	18
	1.1.3 Momentum Representation	24
	1.1.4 Heisenberg Representation	28
	1.1.5 Hartree-Fock Theory	29
1.2	The Fock Space	33
	1.2.1 Occupation Number Representation	35
	1.2.2 Coherent Bosonic States	38
	1.2.3 Grassmann Numbers	42
	1.2.4 Coherent Fermionic States	44
1.3	Field Quantization	47
	1.3.1 Locality	49
	1.3.2 Superselection	50
2	**Macroscopic Quantum Systems**	**55**
2.1	Thermodynamic Limit	55
2.2	Pure and Mixed Quantum States	59
2.3	Thermodynamic States	64
2.4	Stationary States	67
3	**Quasi-Stationary Excitations**	**72**
3.1	Pictures	72
3.2	Time-dependent Green's Functions	75
3.3	Non-Interacting Particles	78
3.4	Interacting Systems	81
3.5	Density of States	85
3.6	Collective Excitations	87
3.7	Non-Zero Temperatures	90

4	**Model Hamiltonians**	**95**
	4.1 Non-Interacting Particles, $T = 0$	95
	4.2 Non-Interacting Particles, $T > 0$	100
	4.3 BCS Theory	104
	4.4 Weakly Interacting Bose Gas	107
	4.5 Discussion of the Model Approach	110

5	**Quasi-Particles**	**114**
	5.1 Landau's Quasi-Particle Conception	115
	5.2 Quasi-Particles in a Crystalline Solid	118
	5.3 Green's Function of Bloch Electrons	121
	5.4 Green's Function of Lattice Phonons	125
	5.5 Disordered Systems	128
	5.6 Frontiers of the Quasi-Particle Picture	131

6	**The Nature of the Vacuum**	**134**
	6.1 Spontaneous Symmetry Breaking	134
	6.2 Gauge Symmetry	137
	6.3 Anomalous Mean Values	139
	6.4 Goldstone and Higgs Modes	143

7	**What Matter Consists of**	**146**
	7.1 The Hierarchy of Hamiltonians	147
	7.2 'From the First Principles'	148

8	**Epilogue: The Unifying Picture of Physics**	**151**

Appendices		**154**
A I	Self-energy operator	154
A II	Density functional theory	155

Bibliography	**160**

Index	**170**

Introduction

The idea of quasi-particles has turned out to be one of the most profound and fundamental ideas in the quantum theory of condensed matter. In terms of phonons, magnons, excitons, conduction electrons and holes and so on, the whole of experimental data and theoretical predictions related to the most various solids under the most different conditions is described. In recent years, a whole zoo of new quasi-particles emerged as composite fermions, spinons, holons, orbitons, phasons, and others. The fertility of solid state research to give birth to new quasi-particles seems not to diminish after five decades.

There are different understandings of quasi-particles. To the one, the quasi-particles serve as a simple approach replacing the 'true' theory by the primitive one-particle approximation. To the other it is the 'remedy from all diseases': the quasi-particles (in their opinion) are capable of describing everything that is going on in a piece of condensed matter, thereby, if something is not yet described by them, it certainly will be in future... One of the goals of the present text is trying to show the true place of quasi-particles (and of particles too) in modern physics.

The macroscopicity (the practically infinite number of microscopic degrees of freedom) of any solid not only introduces specific computational techniques—it also forms the fundamental ideas. So, while for micro-objects (atoms, ions, small molecules) the basic task consists of determining their energy spectra, i.e. the set of *stationary states*, for a macroscopic, condensed system the basic task consists of defining and determining the *quasi-stationary states* describing quasi-particles which, with rare exceptions, have finite lifetimes. Here, a consequent computational way of mathematically extracting the quasi-particles literally from the 'first principles' does not exist. Strictly speaking, even the corresponding problem cannot be formulated, for up to now we do not know the true elementary beings matter consists of (if there is at all such a last pre-matter). Also not going into such theoretical depths and assuming that any solid consists of nuclei and electrons, it is not possible by means of consequent transformations[1] to 'convert' the electro-dynamical Hamiltonian (or even the Coulomb one) of nuclei and electrons into the solid state Hamiltonian with its zoo of quasi-particles (phonons, magnons,

---

[1] rigorous or justified at every step

excitons, conduction electrons and holes, ...). On this way, experiment must be consulted and phenomenology is to be introduced in the theoretical description.

The solid state theory contains this phenomenological level of description not only as a result of reasoning (from a microscopic picture macroscopic relations are deduced), but also as an initial prerequisite. Relying on experimentally observed data, we classify all sorts of condensed matter distinguishing their structures (crystals, quasi-crystals, amorphous solids, liquids), their electric and magnetic properties (conductors, insulators, dia-, para-, ferromagnets), indicating their chemical composition (are they most easily disintegrated into molecules, or atoms, or ions and, eventually, electrons), and, using this knowledge, we try to understand which elementary (comprehensible, experimentally traceable) motions the 'particles' (structural units) of this or the other piece of condensed matter perform. And this kind of reasoning does not go without on the way comparing the results of the theoretical analysis to experimental data, which latter—and this is the main point of our argument—cannot be treated in another way than visualizing that *the elementary motions in condensed matter, though being collectively carried by all its structural units* (molecules, atoms, ions, electrons), *nevertheless display single-particle character*, described in terms suited for the description of the motion of a single quantum particle.

The logical complication in formulating the quasi-particle notion stems from the fact that the usual phrase 'macro... through micro...'(the understanding of macro-objects on the basis of their microstructure) unavoidably needs a closure: the properties of a solid are explained in terms of quasi-particles, but the quasi-particles cannot be described without referring to the solid as a whole. (For instance, thermal expansion of a crystal is explained in terms of phonon-phonon interaction, but the phonons are characterized by quasi-momenta who's definition depends on the lattice constants.) It might appear to the reader that we caught ourselves in a logical cycle. This is of course not the case: fixing the level of description (as will be seen below) allows to introduce the quasi-particles in a logical consistent way and, at the same time, to indicate the limits of applicability of the introduced terms.

For a physicist far away from the field of condensed matter physics, quasi-particles and all constructions connected with them might appear as part of applied physics feeding technology with its results (e.g. 'band structure engineering'). Not denying this important role of quantum physics of condensed matter (especially after the discovery of high temperature superconductivity this would at least not be wise), we intentionally concentrate on the basic principles. Papers, reviews, books, devoted to the description of solids and quantum liquids exist in vast number. We want to shed light on how the quasi-particles appear as an indispensable element of the description of matter. For the understanding of condensed matter physics they are as fundamental and indispensable (and as insufficient) as particles are for the understanding of high energy physics.

In the following chapter of the present text, the technical tools and the fundamental notions for a mathematical description of a quantum many-body system are provided. In Chapter 2, the implications of the thermodynamic limit and of macroscopicity are explained. The apparatus of Green's functions for the description of quasi-particle and of collective excitations is introduced in Chapter 3. Chapter 4 transforms the many-body Hamiltonians of sufficiently simple cases into model Hamiltonians acting in the appropriate Fock spaces. Then, in Chapter 5 the most important quasi-particles of Solid State Theory, the Bloch electrons and the lattice phonons, are considered in more detail. The appropriate Fock space of a solid (at a certain temperature) is determined by the nature of its symmetry-broken vacuum. This point is analyzed in Chapter 6. Finally, Chapter 7 reviews the hierarchy of Hamiltonians if one goes down with the energy scale, and it explains the common phrase 'from the first principles' in Solid State Theory. The main text is closed by an epilogue, Chapter 8, which stresses the close relation between Solid State Theory and Elementary Particle Theory and highlights the unified picture of physics.

This text is written to be lecture notes. Hence, citations and the bibliography at the end are by no means complete and do not document priority of contributions to the field. Nevertheless, besides reviews, monographs and textbooks, also a number of seminal and of early papers is cited, since it is the firm conviction of the present author that the study of important original texts is indispensable for a serious student.

Recommended textbooks on the subject:
(See Bibliography at the end of this text for the complete references.)

[Abrikosov *et al.*, 1975]
[Fetter and Walecka, 1971]
[Fradkin, 1991]
[Kadanoff and Baym, 1989]
[Landau and Lifshits, 1980b]
[Sewell, 1986].

A recommended classic on Relativistic Quantum Field Theory is:

[Itzykson and Zuber, 1980].

Classics in Solid State Theory:

[Ashcroft and Mermin, 1976]
[Madelung, 1978].

1 Technical Tools

Before entering the study of condensed matter physics, in this chapter some compendium of important parts of the mathematical language in quantum theory is provided in order to make this text to a reasonable extent self-contained. The reader is assumed to have an idea of the notion of Hilbert space.

1.1 Résumé on N-Particle Quantum States

A short review over the basic representations of the N-particle quantum state is presented. This comprises the Schrödinger representation, the momentum representation, and the Heisenberg representation with respect to any orthonormal orbital basis. The section is closed with a brief sketch of Hartree-Fock theory.

The general link between all representations of quantum mechanics is the abstract Hilbert space of pure quantum states $|\Psi\rangle$ normalized to unity,

$$\langle \Psi | \Psi \rangle = 1, \tag{1.1}$$

in which Hermitian operators $\hat{A} = \hat{A}^\dagger$ represent observables A,[2] so that the real expectation value $\langle A \rangle$ of the observable A in the quantum state $|\Psi\rangle$ is

$$\langle A \rangle = \langle \Psi | \hat{A} | \Psi \rangle. \tag{1.2}$$

The quantum fluctuation ΔA of the observable A is

$$\Delta A = \langle (A - \langle A \rangle)^2 \rangle^{1/2} = (\langle \Psi | \hat{A}^2 | \Psi \rangle - \langle \Psi | \hat{A} | \Psi \rangle^2)^{1/2}, \tag{1.3}$$

that is, in the same pure quantum state $|\Psi\rangle$, the results of measuring A may scatter with a spread ΔA around $\langle A \rangle$.

The fluctuations vanish in an eigenstate Ψ_a of the observable A,

$$\hat{A} | \Psi_a \rangle = | \Psi_a \rangle a \tag{1.4}$$

---

[2] From a rigorous point of view this would need a specification in case of unbounded operators, which we may disregard here.

with the eigenvalue a being a possible result of measuring A.[3] If a general pure quantum state $|\Psi\rangle$ is realized, the result a of measuring A appears with probability

$$p(a) = |\langle\Psi_a|\Psi\rangle|^2 = \langle\Psi|\Psi_a\rangle\langle\Psi_a|\Psi\rangle. \tag{1.5}$$

Since the sum of the probabilities $p(a)$ over all possible results a of the measurement must be unity in every state $|\Psi\rangle$, the eigenstates $|\Psi_a\rangle$ must form a complete set,

$$\sum_a |\Psi_a\rangle\langle\Psi_a| = 1, \tag{1.6}$$

where the right side means the identity operator in the physical Hilbert space.[4]

The dynamics of a quantum system is governed by its Hamiltonian $\hat{H}$ and is either, in the *Schrödinger picture*, described by time-dependent states

$$-\frac{\hbar}{i}\frac{\partial}{\partial t}|\Psi(t)\rangle = \hat{H}|\Psi(t)\rangle \tag{1.7}$$

and time-independent operators $\hat{A}$, or, in the *Heisenberg picture*, by time-dependent operators

$$\frac{\hbar}{i}\frac{\partial}{\partial t}\hat{A}(t) = \hat{H}\hat{A}(t) - \hat{A}(t)\hat{H} \tag{1.8}$$

and time-independent states $|\Psi\rangle$. All matrix elements $\langle\Psi|\hat{A}|\Psi'\rangle$ of observables have the same time dependence in both cases (exercise).

---

[3] For a bounded operator the set of eigenvalues a is bounded: $\sup|a| = \|\hat{A}\|$, where $\|\hat{A}\|$ is the norm of the operator.

[4] In most cases, this sum is an infinite series, that is, the physical Hilbert space is infinite dimensional. Moreover, in case of a continuous spectrum of $\hat{A}$, Eq. (1.6) is to be understood as an abbreviation of

$$\sum_a |\Psi_a\rangle\langle\Psi_a| + \int d\alpha\, |\Psi_\alpha\rangle\langle\Psi_\alpha| = 1,$$

where the sum is over the discrete part of the spectrum of $\hat{A}$ and the integral is over the (absolute) continuous part. The ortho-normalization of both the discrete eigenstates and the continuum "eigenstates" is

$$\langle\Psi_a|\Psi_{a'}\rangle = \delta_{aa'}, \quad \langle\Psi_a|\Psi_\alpha\rangle = 0, \quad \langle\Psi_\alpha|\Psi_{\alpha'}\rangle = \delta(\alpha - \alpha').$$

For details see [Bohm, 1993]. An alternative is the use of an abstract spectral measure [von Neumann, 1955].

1.1 Résumé on N-Particle Quantum States

For brevity of notation, the same symbol $\hat{A}$ will be used for the operator of the observable A in all representations. For an N-particle system with N conserved the Hilbert space is complex. The complex conjugate to the complex number z will be denoted by z^*.

1.1.1 Schrödinger Representation

This is the spatial wavefunction representation. As this text is mainly concerned with spin zero and spin 1/2 particles, we outline here the spin 1/2 case and mention ocassionally the general spin S case for arbitrary integer or half-integer S. Formally[5] the eigenstates of the particle position operator $\hat{r}$ are introduced as

$$\hat{r}\,|r\rangle = |r\rangle\,r. \tag{1.9}$$

For the spin 1/2 operator $(\hbar/2)\hat{\boldsymbol{\sigma}}$ the 2×2 Pauli matrices

$$\hat{\sigma}_x = \begin{pmatrix} 0 & 1 \\ 1 & 0 \end{pmatrix}, \quad \hat{\sigma}_y = \begin{pmatrix} 0 & -i \\ i & 0 \end{pmatrix}, \quad \hat{\sigma}_z = \begin{pmatrix} 1 & 0 \\ 0 & -1 \end{pmatrix}. \tag{1.10}$$

are introduced together with the eigenstates $|s\rangle$ of the spin projection with respect to some given quantization axis z of the particles

$$\hat{\sigma}_z\,|s\rangle = |s\rangle\,s, \quad s = \pm 1. \tag{1.11}$$

In combination with the position eigenstates they form a basis of unit vectors in the Hilbert space of one-particle quantum states. A combined variable

$$x \stackrel{\text{def}}{=} (\boldsymbol{r}, s), \quad |x\rangle = |\boldsymbol{r}\rangle|s\rangle = |\boldsymbol{r}s\rangle, \quad \int dx \stackrel{\text{def}}{=} \sum_s \int d^3r \tag{1.12}$$

will be used for both position and spin of a particle. In case of spinless particles, $s = 0$ can be omitted and x and $\boldsymbol{r}$ may be considered synonyms.

The N-particle quantum state is now represented by a (spinor-)wavefunction

$$\Psi(x_1 \ldots x_N) = \langle x_1 \ldots x_N | \Psi \rangle = \langle \boldsymbol{r}_1 s_1 \ldots \boldsymbol{r}_N s_N | \Psi \rangle. \tag{1.13}$$

For spin-S particles, the spin variable s_i runs over $2S+1$ values. For one spin-half particle, e.g., the spinor part of the wavefunction (for fixed $\boldsymbol{r}$),

$$\chi(s) = \langle s|\chi\rangle = \begin{pmatrix} \chi_+ \\ \chi_- \end{pmatrix}, \tag{1.14}$$

---

[5] Cf. the previous footnote.

consists of two complex numbers χ_+ and χ_-, forming the components of a $SU(2)$-spinor. This latter statement means that a certain linear transformation of those two components is linked to every spatial rotation of the $\boldsymbol{r}$-space (see next subsection).

The eigenstates of $\hat{\sigma}_z$,

$$\chi^+(s) = \langle s|\chi^+\rangle = \begin{pmatrix} 1 \\ 0 \end{pmatrix}, \quad \chi^-(s) = \begin{pmatrix} 0 \\ 1 \end{pmatrix}, \tag{1.15}$$

form a complete set for the s-dependence at a given space-point $\boldsymbol{r}$:

$$\langle \chi^+|\chi\rangle = \sum_s \langle \chi^+|s\rangle\langle s|\chi\rangle = (1\ 0)\begin{pmatrix} \chi_+ \\ \chi_- \end{pmatrix} = \chi_+, \quad \langle \chi^-|\chi\rangle = \chi_-. \tag{1.16}$$

The full wavefunction of a spin-half particle is given by

$$\phi(x) = \phi(\boldsymbol{r}s) = \begin{pmatrix} \phi_+(\boldsymbol{r}) \\ \phi_-(\boldsymbol{r}) \end{pmatrix}. \tag{1.17}$$

It is called a spin-orbital.

For fermions (half-integer spin), only wavefunctions, which are antisymmetric with respect to particle exchange, are admissible:

$$\Psi(x_1 \ldots x_i \ldots x_k \ldots x_N) = -\Psi(x_1 \ldots x_k \ldots x_i \ldots x_N). \tag{1.18}$$

In the non-interacting case, Slater determinants

$$\Phi_L(x_1 \ldots x_N) = \frac{1}{\sqrt{N!}} \det \|\phi_{l_i}(x_k)\| \tag{1.19}$$

of single-particle wavefunctions $\phi_{l_i}(x_k)$ (spin-orbitals) out of some given set $\{\phi_l\}$ are appropriate. The determinant (1.19) can be non-zero only if the orbitals ϕ_{l_i} are linear independent, it is normalized if the orbitals are orthonormal. Furthermore, if the ϕ_{l_i} may be written as $\phi_{l_i} = \phi'_{l_i} + \phi''_{l_i}$, where $\langle \phi'_{l_i}|\phi'_{l_j}\rangle \sim \delta_{ij}$, and the ϕ''_{l_i} are linear dependent on the ϕ'_{l_j}, $j \neq i$, then the value of the determinant depends on the orthogonal to each other parts ϕ'_{l_i} only. These statements following from simple determinant rules comprise Pauli's exclusion principle for fermions. The subscript L of Φ_L denotes an orbital configuration $L \stackrel{\text{def}}{=} (l_1 \ldots l_N)$. For a given fixed *complete* set of spin-orbitals $\phi_l(x)$, i.e. for a set with the property

$$\sum_l \phi_l(x)\phi_l^*(x') = \delta(x - x') = \delta_{ss'}\delta(\boldsymbol{r} - \boldsymbol{r}'), \tag{1.20}$$

the Slater determinants for all possible orbital configurations span the antisymmetric sector of the N-particle Hilbert space which is the fermionic N-particle Hilbert space. In particular, the general state (1.18) may be expanded according to

$$\Psi(x_1 \ldots x_N) = \sum_L C_L \Phi_L(x_1 \ldots x_N) \qquad (1.21)$$

('configuration interaction').

Bosonic (integer, in particular zero spin) wavefunctions must be symmetric with respect to particle exchange:

$$\Psi(x_1 \ldots x_i \ldots x_k \ldots x_N) = \Psi(x_1 \ldots x_k \ldots x_i \ldots x_N). \qquad (1.22)$$

The corresponding symmetric sector of the N-particle Hilbert space is formed by the product states (so-called permanents)

$$\Phi_L(x_1 \ldots x_N) = \mathcal{N} \sum_{\mathcal{P}} \prod_i \phi_{l_{\mathcal{P}i}}(x_i), \qquad (1.23)$$

where $\mathcal{N}$ is a normalization factor, and $\mathcal{P}$ means a permutation of the subscripts $12 \ldots N$ into $\mathcal{P}1\mathcal{P}2\ldots\mathcal{P}N$. The subscripts l_i, $i = 1\ldots N$, need not be different from each other in this case. Particularly all ϕ_{l_i} might be equal to each other (Bose condensation).

If the ϕ_{l_i} are again taken out of some given orthonormal set $\{\phi_k\}$ and if the orbital ϕ_k appears n_k times in the permanent Φ_L, then it is easily shown that

$$\mathcal{N} = \left(N! \prod_k n_k! \right)^{-1/2}, \quad \sum_k n_k = N. \qquad (1.24)$$

(Exercise.)

For both fermionic and bosonic systems, the probability density of a given configuration $(x_1 \ldots x_N)$,

$$p(x_1 \ldots x_N) = \Psi^*(x_1 \ldots x_N)\Psi(x_1 \ldots x_N), \qquad (1.25)$$

is independent of particle exchange.

A local operator in position space like a local potential acts on the wavefunction by multiplication with a position dependent function like the position operator itself. A spin dependent operator like a magnetic field coupling to the magnetic spin moment or the dipolar interaction of two magnetic spin moments acts by multiplication with a 2×2 spin matrix for each spin $1/2$ particle. Thus, the Hamiltonian of N particles moving in the (possibly spin-dependent) external potential $v_{ss'}(\boldsymbol{r})$ and interacting instantaneously via a (possibly spin-dependent) pair potential $w_{s_1 s_1', s_2 s_2'}(\boldsymbol{r}_1, \boldsymbol{r}_2)$ reads

$$\hat{H} = -\frac{\hbar^2}{2m} \sum_{i=1}^{N} \nabla_i^2 + \sum_{i=1}^{N} v_{s_i s_i'}(\boldsymbol{r}_i) + \frac{1}{2} \sum_{i \neq j}^{N} w_{s_i s_i', s_j s_j'}(\boldsymbol{r}_i, \boldsymbol{r}_j). \tag{1.26}$$

The connection with the abstract Hilbert space representation is given by

$$\sum_{x_1'' \ldots x_N''} \langle x_1 \ldots x_N | \hat{H} | x_1'' \ldots x_N'' \rangle \langle x_1'' \ldots x_N'' | \Psi \rangle =$$

$$= \left(-\frac{\hbar^2}{2m} \sum_{i=1}^{N} \sum_{s_i'} \delta_{s_i s_i'} \nabla_i^2 + \sum_{i=1}^{N} \sum_{s_i'} v_{s_i s_i'}(\boldsymbol{r}_i) + \right.$$

$$\left. + \frac{1}{2} \sum_{i \neq j}^{N} \sum_{s_i' s_j'} w_{s_i s_i', s_j s_j'}(\boldsymbol{r}_i, \boldsymbol{r}_j) \right) \langle x_1' \ldots x_N' | \Psi \rangle, \tag{1.27}$$

where the sum over the $x_i'' = (\boldsymbol{r}_i', s_i')$ runs over the fermionic (bosonic) sector only, $x_i' = (\boldsymbol{r}_i, s_i')$, and $\hat{H}$ on the left means the abstract Hilbert space operator.

1.1.2 Spin

By direct calculation one finds the algebraic properties of the Pauli spin matrices (1.10):

$$\hat{\sigma}_\alpha \hat{\sigma}_\beta - \hat{\sigma}_\beta \hat{\sigma}_\alpha = 2i \varepsilon_{\alpha\beta\gamma} \hat{\sigma}_\gamma, \qquad \hat{\sigma}_\alpha \hat{\sigma}_\beta + \hat{\sigma}_\beta \hat{\sigma}_\alpha = 2 \delta_{\alpha\beta} 1_2. \tag{1.28}$$

Here, α, β, γ are Cartesian subscripts, $\varepsilon_{\alpha\beta\gamma}$ is the totally antisymmetric form (summation over the repeated subscript γ understood), and 1_2 means the 2×2 unit matrix. The first relation has the well known form of the commutation relations of the components of angular momentum. Both relations combine into[6]

$$\hat{\sigma}_\alpha \hat{\sigma}_\beta = \delta_{\alpha\beta} 1_2 + i \varepsilon_{\alpha\beta\gamma} \hat{\sigma}_\gamma. \tag{1.29}$$

---

[6]With these algebraic relations, the combination $H = (a, b, c, d) = a 1_2 + b i \hat{\sigma}_z + c i \hat{\sigma}_y + d i \hat{\sigma}_x$ of four real numbers a, b, c, d acquires the properties of a quaternion.

1.1 Résumé on N-Particle Quantum States

In addition we note the properties

$$\hat{\sigma}_\alpha^\dagger = \hat{\sigma}_\alpha, \qquad \text{tr}\,\hat{\sigma}_\alpha = 0, \qquad \text{tr}\,(\hat{\sigma}_\alpha \hat{\sigma}_\beta) = 2\delta_{\alpha\beta}. \tag{1.30}$$

Here, tr means the trace of the matrix. The first two of these relations are obvious, the last one follows directly from (1.29).

If we treat the Pauli matrices (1.10) as the three components of a three-vector $\hat{\boldsymbol{\sigma}} = (\hat{\sigma}_x, \hat{\sigma}_y, \hat{\sigma}_z)$, this vector provides a one-to-one map of ordinary vectors $\boldsymbol{r}$ onto Hermitian, traceless 2×2 matrices:

$$\boldsymbol{r} \;\mapsto\; \boldsymbol{r}\cdot\hat{\boldsymbol{\sigma}} = P = \begin{pmatrix} z & x - iy \\ x + iy & -z \end{pmatrix} \;\mapsto\; \frac{1}{2}\text{tr}\,(P\hat{\boldsymbol{\sigma}}) = \boldsymbol{r}. \tag{1.31}$$

Note that P describes a spatial position here, not a spin structure. From the structure of P we read off

$$\det P = -r^2. \tag{1.32}$$

A direct consequence of (1.29) is

$$P_1 P_2 = (\boldsymbol{r}_1 \cdot \hat{\boldsymbol{\sigma}})(\hat{\boldsymbol{\sigma}} \cdot \boldsymbol{r}_2) = \boldsymbol{r}_1 \cdot \boldsymbol{r}_2 1_2 + i(\boldsymbol{r}_1 \times \boldsymbol{r}_2)\cdot \hat{\boldsymbol{\sigma}}, \tag{1.33}$$

that is, the product of P-matrices corresponds to covariant vector operations of the $\boldsymbol{r}$-vectors.

Consider now an $SO(3)$-rotation of the $\boldsymbol{r}$-space:

$$\boldsymbol{r}' = O\boldsymbol{r}, \qquad O^T = O^{-1}, \qquad \det O = 1. \tag{1.34}$$

We want to find the corresponding transformation of the P-matrices. Clearly it must be linear in P and it must preserve the Hermitian property, hence $P' = UPU^\dagger$ with a certain 2×2 matrix U. From (1.33) it follows $(P_1 P_2)' = P_1' P_2'$, and hence $U^\dagger = U^{-1}$, that is, U must be unitary. P' is not affected by a change $U \to e^{i\alpha} U$ with α real. By choosing $\alpha = -\arg(\det U)/2$ we can always have $\det U = +1$ for all U. Hence we found a mapping

$$SO(3) \ni O \quad \mapsto \quad U \in SU(2). \tag{1.35}$$

This mapping is in fact a local isomorphism of Lie groups. The rotation O may be parametrized by Euler angles: a rotation by φ around the z-axis followed by a

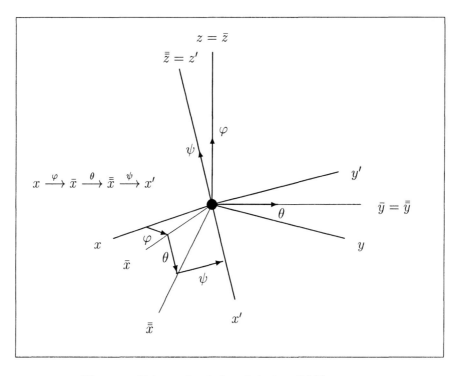

Figure 1: Euler angles ϕ, θ, and ψ of an $SO(3)$-rotation

rotation by θ around the new y-axis followed by a rotation by ψ around the new z-axis (see Fig.1). The corresponding unitary transformation U is

$$U = e^{\frac{i}{2}\psi\hat{\sigma}_z} e^{\frac{i}{2}\theta\hat{\sigma}_y} e^{\frac{i}{2}\varphi\hat{\sigma}_z}, \tag{1.36}$$

where the exponentials are to be understood as power series:

$$\begin{aligned} e^{\frac{i}{2}\varphi\hat{\sigma}_z} &= 1 + i\frac{\varphi}{2}\hat{\sigma}_z - \frac{1}{2!}\left(\frac{\varphi}{2}\right)^2 - \frac{i}{3!}\left(\frac{\varphi}{2}\right)^3 \hat{\sigma}_z + \cdots \\ &= \begin{pmatrix} \cos(\varphi/2) + i\sin(\varphi/2) & 0 \\ 0 & \cos(\varphi/2) - i\sin(\varphi/2) \end{pmatrix}, \end{aligned} \tag{1.37}$$

1.1 Résumé on N-Particle Quantum States

$$\begin{aligned} e^{\frac{i}{2}\theta\hat{\sigma}_y} &= 1 + i\frac{\theta}{2}\hat{\sigma}_y - \frac{1}{2!}\left(\frac{\theta}{2}\right)^2 - \frac{i}{3!}\left(\frac{\theta}{2}\right)^3 \hat{\sigma}_y + \cdots \\ &= \begin{pmatrix} \cos(\theta/2) & \sin(\theta/2) \\ -\sin(\theta/2) & \cos(\theta/2) \end{pmatrix}. \end{aligned} \quad (1.38)$$

Hence,

$$U = \begin{pmatrix} \cos(\theta/2)e^{i(\psi+\phi)/2} & \sin(\theta/2)e^{i(\psi-\phi)/2} \\ -\sin(\theta/2)e^{-i(\psi-\phi)/2} & \cos(\theta/2)e^{-i(\psi+\phi)/2} \end{pmatrix}. \quad (1.39)$$

With these formulas it is straightforwardly seen that, now applied to the spin operator proper, for instance,

$$O_\varphi \hat{\boldsymbol{\sigma}} = (\cos\varphi\,\hat{\sigma}_x - \sin\varphi\,\hat{\sigma}_y,\ \sin\varphi\,\hat{\sigma}_x + \cos\varphi\,\hat{\sigma}_y,\ \hat{\sigma}_z) = U_\varphi \hat{\boldsymbol{\sigma}} U_\varphi^\dagger, \quad (1.40)$$

where O_φ is an $SO(3)$ rotation by φ around the z-axis, acting on the components of the three-vector $\hat{\boldsymbol{\sigma}}$ whose components are 2×2 spin matrices, while U_φ is the $SU(2)$ transformation (1.37), acting on a 2×2 spin matrix whose components are three-vectors. Analogous relations are straightforwardly obtained for the other elementary Euler rotations and hence hold true for a general $SO(3)$ rotation and its $SU(2)$ counterpart (1.36).

Since $\hat{\boldsymbol{r}}$ and $\hat{\boldsymbol{\sigma}}$ are independent dynamical variables of an orbital, we must distinguish between the $SO(3)$ rotation O of the position space as in (1.34) and the $SO(3)$ rotation O of the spin space as in (1.40). Since the latter can always be expressed through the $SU(2)$ transformation U of (1.40), we will not use it any more, and in what follows, O means always a rotation of the $\boldsymbol{r}$-space while a rotation of the spin space is expressed by the corresponding transformation U.

If the Hamiltonian does not explicitly depend on the spin, then to every spinor solution ϕ of the Schrödinger equation, $\phi' = U\phi$ is again a solution, with the same energy in the stationary case. Analogously, if the potential is spherically symmetric, to every solution $\phi(\boldsymbol{r})$, $\phi'(\boldsymbol{r}) = \phi(O^{-1}\boldsymbol{r})$ is again a solution. If, however, the Hamiltonian contains a term proportional to $\hat{\boldsymbol{l}}\cdot\hat{\boldsymbol{\sigma}}$ where $\hat{\boldsymbol{l}}$ is a vector of the position space (orbital angular momentum operator), that is, if spin-orbit coupling exists, then the only remaining symmetry transformations are simultaneous rotations of position and spin space by the same Euler angles:

$$\phi'(\boldsymbol{r}s) = \sum_{s'} U_{ss'}\phi(O^{-1}\boldsymbol{r}s'). \quad (1.41)$$

In other words, spin-orbit coupling sets up a one-one correspondence between directions in position space and directions in spin space. The same is established by an interaction term proportional to $\boldsymbol{B} \cdot \hat{\boldsymbol{\sigma}}$, where $\boldsymbol{B}$ is an applied magnetic field (which is due to electric currents in position space). This means that a magnetic spin moment is always based on this one-one correspondence between directions in position and spin spaces. In the writings of the orbital transformations we have used the notation of the transformation law of a tensor field (active transformation of the components and passive transformation of the function variable).

In the case of a many-particle state (1.18) one introduces a spin operator for each particle so that the spin operator of the k-th particle acts according to

$$(\overset{k}{\hat{\boldsymbol{\sigma}}}\Psi)(\ldots \boldsymbol{r}_k s_k \ldots) = \sum_{s'_k} \boldsymbol{\sigma}_{s_k s'_k} \Psi(\ldots \boldsymbol{r}_k s'_k \ldots). \tag{1.42}$$

$((\sigma_\alpha)_{ss'}$ is the matrix element of the Pauli matrix $\hat{\sigma}_\alpha$; in this notation, for fixed s and s', $\boldsymbol{\sigma}_{ss'} = ((\sigma_x)_{ss'}, (\sigma_y)_{ss'}, (\sigma_z)_{ss'})$ is again a three-vector in spin space.) The commutation relations for those spin operators are obviously

$$\overset{k}{\hat{\sigma}}_\alpha \overset{l}{\hat{\sigma}}_\beta - \overset{l}{\hat{\sigma}}_\beta \overset{k}{\hat{\sigma}}_\alpha = 2i\delta_{kl}\varepsilon_{\alpha\beta\gamma}\overset{k}{\hat{\sigma}}_\gamma, \tag{1.43}$$

since spin operators for different particles operate on different variables, they commute.

Since the spin operators for the Cartesian components of the spin do not commute, these components cannot be measured simultaneously. Nevertheless, in any state (1.17) the expectation value of the vector spin is well defined according to

$$\boldsymbol{S} = \frac{\hbar}{2}\langle \phi|\hat{\boldsymbol{\sigma}}|\phi\rangle = \frac{\hbar}{2}\sum_{ss'}\int d^3r\, \phi^*(\boldsymbol{r}s)\boldsymbol{\sigma}_{ss'}\phi(\boldsymbol{r}s'). \tag{1.44}$$

The components of this vector fluctuate according to the general rule (1.3). In this respect there is a total analogy to the situation with an angular momentum. However, unlike the case of the angular momentum which is an intrinsically nonlocal entity, there is even a local density of the vector spin expressed through the spin density matrix

$$n_{ss'}(\boldsymbol{r}) = \phi(\boldsymbol{r}s)\phi^*(\boldsymbol{r}s'), \tag{1.45}$$

the trace of which is the particle density:

$$n(\boldsymbol{r}) = \operatorname{tr}\hat{n}(\boldsymbol{r}) = n_{++}(\boldsymbol{r}) + n_{--}(\boldsymbol{r}). \tag{1.46}$$

1.1 Résumé on N-Particle Quantum States

If we define the vector spin density as

$$\boldsymbol{S}(\boldsymbol{r}) = \frac{\hbar}{2}\operatorname{tr}(\hat{n}(\boldsymbol{r})\hat{\boldsymbol{\sigma}}) = \frac{\hbar}{2}\sum_{ss'} n_{s's}(\boldsymbol{r})\boldsymbol{\sigma}_{ss'} = \frac{\hbar}{2}\phi^*(\boldsymbol{r}s)\boldsymbol{\sigma}_{ss'}\phi(\boldsymbol{r}s'), \qquad (1.47)$$

then (1.44) is obtained from (1.47) by integration over the $\boldsymbol{r}$-space. Of course, the existence of this vector spin density follows from the fact, that, unlike the orbital momentum operator, the spin operator commutes with the position operator (1.9). Moreover, the vector directions in this spin vector space are uniquely linked to directions of the position space in case of spin-orbit coupling or an applied magnetic field only.

For the N-particle state (1.18) the spin density matrix is defined as

$$n_{ss'}(\boldsymbol{r}) = N\int dx_2\cdots dx_N\,\Psi(\boldsymbol{r}s, x_2\cdots x_N)\Psi^*(\boldsymbol{r}s', x_2\cdots x_N). \qquad (1.48)$$

Since $\Psi\Psi^*$ is totally symmetric in the particle variables, a summation over all particles can simply be replaced by N times the expression for particle 1. Hence, the connection with the total particle density and the total vector spin density or vector spin is the same as in (1.46), (1.47) and (1.44).

From (1.47) we read off the expressions for the Cartesian components of the vector spin density:

$$\begin{aligned} S_x(\boldsymbol{r}) &= \frac{\hbar}{2}(n_{+-}(\boldsymbol{r}) + n_{-+}(\boldsymbol{r})), \quad S_y(\boldsymbol{r}) = i\frac{\hbar}{2}(n_{+-}(\boldsymbol{r}) - n_{-+}(\boldsymbol{r})), \\ S_z(\boldsymbol{r}) &= \frac{\hbar}{2}(n_{++}(\boldsymbol{r}) - n_{--}(\boldsymbol{r})). \end{aligned} \qquad (1.49)$$

The spin flip amplitudes for rising and lowering the z-component of the spin are

$$\begin{aligned} S_+(\boldsymbol{r}) &= S_x(\boldsymbol{r}) + iS_y(\boldsymbol{r}) = \hbar n_{-+}(\boldsymbol{r}), \\ S_-(\boldsymbol{r}) &= S_x(\boldsymbol{r}) - iS_y(\boldsymbol{r}) = \hbar n_{+-}(\boldsymbol{r}). \end{aligned} \qquad (1.50)$$

(Check the relations $\hat{\sigma}_+\chi^- = 2\chi^+$, $\hat{\sigma}_-\chi^+ = 2\chi^-$ for $\hat{\sigma}_\pm = \hat{\sigma}_x \pm i\hat{\sigma}_y$.) We have written down explicitly the $\boldsymbol{r}$-dependence in all those expressions in order to make clear that they are well defined densities.[7]

---

[7] To measure a density in (non-relativistic) quantum physics amounts to use the observable $\boldsymbol{r}$, which is an idealized observable, in Schrödinger representation given by a δ-function multiplication operator. In truth, measuring a density means projection onto a well localized wave pocket.

1.1.3 Momentum Representation

Here, the main features of the momentum or plane wave representation used in the following text are summarized.[8] Formally it uses the eigenstates of the particle momentum operator

$$\hat{\boldsymbol{p}}|\boldsymbol{k}\rangle = |\boldsymbol{k}\rangle\,\hbar\boldsymbol{k} \tag{1.51}$$

instead of the position vector eigenstates from (1.9) as basis vectors in the Hilbert space of one-particle quantum states.

In Schrödinger representation, the particle momentum operator is given by $\hat{\boldsymbol{p}} = (\hbar/i)\,\nabla$, and the momentum eigenstate (1.51) is represented by a wavefunction

$$\phi_{\boldsymbol{k}}(\boldsymbol{r}) = \langle \boldsymbol{r}|\boldsymbol{k}\rangle = \frac{1}{\sqrt{V}}e^{i\boldsymbol{k}\cdot\boldsymbol{r}}. \tag{1.52}$$

V is the total volume or the normalization volume (often tacitly put equal to unity).

In Solid State Theory it is convenient to use periodic boundary conditions also called Born-von Kármán boundary conditions. They avoid formal mathematical problems with little physical relevance with the continuous spectrum of energy eigenstates and simplify the treatment of the thermodynamic limes. They consist in replacing the infinite position space $\boldsymbol{R}^3$ of the particles by a large torus $\boldsymbol{T}^3$ of volume $V = L^3$ defined by

$$x + L \equiv x, \quad y + L \equiv y, \quad z + L \equiv z, \tag{1.53}$$

where (x, y, z) are the components of the position vector. The meaning of (1.53) is that any function of x, y, z must fulfill the periodicity conditions $f(x + L) = f(x)$, and so on. As is immediately seen from (1.52), this restricts the spectrum of eigenvalues $\boldsymbol{k}$ of (1.51) to the values

$$\boldsymbol{k} = \frac{2\pi}{L}\,(n_x, n_y, n_z) \tag{1.54}$$

with integers n_x, n_y, n_z. The $\boldsymbol{k}$-values (1.54) form a simple cubic mesh in the wavenumber space ($\boldsymbol{k}$-space) with a $\boldsymbol{k}$-space density of states (number of $\boldsymbol{k}$-vectors (1.54) within a unit volume of $\boldsymbol{k}$-space)

$$D(\boldsymbol{k}) = \frac{V}{(2\pi)^3}, \quad \text{i.e.} \quad \sum_{\boldsymbol{k}} \rightarrow \frac{V}{(2\pi)^3}\int d^3k. \tag{1.55}$$

---

[8] Pages 24 to 33 and 35 to 37 contain passages of text from [Eschrig, 2003] which is repeated here in order to make the present text sufficiently self-contained.

1.1 Résumé on N-Particle Quantum States

In the thermodynamic limes $V \to \infty$ one simply has $D(\mathbf{k})/V = (2\pi)^{-3}\sum_s \int d^3k$. Again we introduce a combined variable

$$q \stackrel{\text{def}}{=} (\mathbf{k}, s), \quad \sum_q \stackrel{\text{def}}{=} \sum_s \sum_{\mathbf{k}} = \frac{V}{(2\pi)^3}\sum_s \int d^3k. \tag{1.56}$$

for both momentum and spin of a particle.

In analogy to (1.13), the N-particle quantum state is now represented by a (spinor-)wavefunction in momentum space

$$\Psi(q_1 \ldots q_N) = \langle q_1 \ldots q_N | \Psi \rangle = \langle \mathbf{k}_1 s_1 \ldots \mathbf{k}_N s_N | \Psi \rangle \tag{1.57}$$

expressing the probability amplitude of a particle momentum (and possibly spin) configuration $(q_1 \ldots q_N)$ in complete analogy to (1.25). Everything that was said in the preceding section between (1.9) and (1.25) transfers accordingly to the present representation. Particularly, in the case of fermions, Ψ of (1.57) is totally antisymmetric with respect to permutations of the q_i, and it is totally symmetric in the case of bosons. Its spin dependence is in complete analogy to that of the Schrödinger wavefunction (1.13).

Equation (1.4) reads in momentum representation

$$\sum_{(q'_1 \ldots q'_N)} \langle q_1 \ldots q_N | \hat{A} | q'_1 \ldots q'_N \rangle \langle q'_1 \ldots q'_N | \Psi_a \rangle = \langle q_1 \ldots q_N | \Psi_a \rangle\, a, \tag{1.58}$$

where the summation runs over the physically distinguished states only.

With this rule, it is a bit tedious but not really complicated to cast the Hamiltonian (1.26) into the momentum representation:

$$\langle q_1 \ldots q_N | \hat{H} | q'_1 \ldots q'_N \rangle = \frac{\hbar^2}{2m}\sum_i k_i^2 \prod_j \delta_{q_j q'_j} +$$
$$+ \sum_i v^{s_{\bar{i}} s'_i}_{\mathbf{k}_{\bar{i}} - \mathbf{k}'_i} (\mp 1)^{\bar{\mathcal{P}}} \prod_{j(\neq i)} \delta_{q_{\bar{j}} q'_j} +$$
$$+ \frac{1}{2}\sum_{i \neq j}\left[w^{s_{\bar{i}} s'_i, s_{\bar{j}} s'_j}_{\mathbf{k}_{\bar{i}} - \mathbf{k}'_i, \mathbf{k}_{\bar{j}} - \mathbf{k}'_j} \mp w^{s_{\bar{i}} s'_j, s_{\bar{j}} s'_i}_{\mathbf{k}_{\bar{i}} - \mathbf{k}'_j, \mathbf{k}_{\bar{j}} - \mathbf{k}'_i}\right](\mp 1)^{\bar{\mathcal{P}}} \prod_{k(\neq i,j)} \delta_{q_{\bar{k}} q'_k}. \tag{1.59}$$

Here, $i = \bar{\mathcal{P}}\bar{i}$, and $\bar{\mathcal{P}}$ is a permutation of the subscripts which puts $\bar{i}$ in the position i (and puts $\bar{j}$ in the position j in the last sum) and leaves the order of the remaining subscripts unchanged. There is always at most one permutation $\bar{\mathcal{P}}$ (up to an irrelevant interchange of $\bar{i}$ and $\bar{j}$, and in the bosonic case up to ineffective permutations of identical states) for which the product of Kronecker δ's can be non-zero. For the diagonal matrix element, that is $q_i' = q_i$ for all i, $\bar{\mathcal{P}}$ is the identity and can be omitted.

The Fourier transforms of $v_{ss'}(\mathbf{r})$ and $w_{s_1 s_1', s_2 s_2'}(\mathbf{r}_1, \mathbf{r}_2)$ are given by

$$v_{\mathbf{k}}^{ss'} = \frac{1}{V} \int d^3 r \, v_{ss'}(\mathbf{r}) e^{-i\mathbf{k} \cdot \mathbf{r}} \tag{1.60}$$

and

$$w_{\mathbf{k}_1, \mathbf{k}_2}^{s_1 s_1', s_2 s_2'} = \frac{1}{V^2} \int d^3 r_1 d^3 r_2 \, w_{s_1 s_1', s_2 s_2'}(\mathbf{r}_1, \mathbf{r}_2) e^{-i\mathbf{k}_1 \cdot \mathbf{r}_1 - i\mathbf{k}_2 \cdot \mathbf{r}_2}. \tag{1.61}$$

The latter expression simplifies further, if the pair interaction depends on the particle distance only.

The elementary processes corresponding to the terms of (1.59) are depicted in Fig.2. The first sum of (1.59) is over the kinetic energies of the particles in their momentum eigenstates (1.51, 1.52). Since the momentum operator of single particles commutes with the kinetic energy operator of the system, this part is diagonal in momentum representation, which is formally expressed by the j-product over Kronecker symbols $\delta_{q_j q_j'}$. The next sum contains the individual interaction events of particles with the external potential $v_{ss'}(\mathbf{r})$. The amplitude of this interaction process is given by the Fourier transform of the potential. Since the interaction of each particle with the external field is assumed to be independent of the other particles (the corresponding term of $\hat{H}$ is assumed to be a sum over individual items $v_{s_l s_l'}(\mathbf{r}_l)$ in (1.26)), all remaining particle states $j \neq i$ are kept unchanged in an interaction event in which one particle makes a transition from the state q_i' to the state $q_{\bar{i}}$ (from the state q_3' to the state $q_{\bar{3}} = q_2$ in the above considered example). The constancy of the remaining particle states is again expressed by the j-product. Finally, the classical imagination of an elementary event of pair interaction is that particles in states $\mathbf{k}_i' s_i'$ and $\mathbf{k}_j' s_j'$ collide and are scattered into states $\mathbf{k}_{\bar{i}} s_{\bar{i}}$ and $\mathbf{k}_{\bar{j}} s_{\bar{j}}$. Quantum-mechanically, one cannot decide which of the particles, formerly in states q_i', q_j', is afterwards in the state $q_{\bar{i}}$ and which one is in the state $q_{\bar{j}}$. This leads to the second term in (1.59), the exchange term with $q_{\bar{i}}$ and $q_{\bar{j}}$ reversed. For a fermionic system, the exchange term appears with a negative sign, and for a bosonic system, with a positive sign.

1.1 Résumé on N-Particle Quantum States

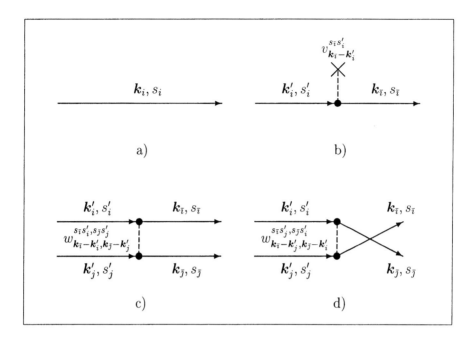

Figure 2: Elementary events corresponding to the terms contained in (1.59):
a) Propagation of a particle with (conserved) momentum k_i and spin s_i, and having kinetic energy $\hbar^2 k_i^2/2m$.
b) Scattering event of a particle having initially momentum k'_i and spin s'_i, on the potential $v_{ss'}(r)$.
c) Direct interaction event of a pair of particles having initially momenta k'_i and k'_j and spins s'_i and s'_j.
d) Exchange interaction event of a pair of particles.

An important reference system is formed by interaction-free fermions ($w = 0$) in a constant external potential $v_{ss'}(r) = 0$: the homogeneous interaction-free fermion gas with the Hamiltonian $\hat{H}_f$:

$$\langle q_1 \ldots q_N | \hat{H}_f | q'_1 \ldots q'_N \rangle = \frac{\hbar^2}{2m} \sum_i k_i^2 \prod_j \delta_{q_j q'_j}. \tag{1.62}$$

Its ground state is a Slater determinant of plane waves so that the sum $\sum_{i=1}^{N} k_i^2$ is minimum. This is obviously the case if all k_i lie inside a sphere of radius k_f

determined by

$$N = \sum_{q_i}^{k_i \leq k_f} 1 = (2S+1) \frac{V}{(2\pi)^3} \int_{k \leq k_f} d^3k, \qquad (1.63)$$

where the factor $(2S+1)$ in front of the last expression comes from summation over the spin values for each $\boldsymbol{k}$. Hence

$$\frac{N}{V} = n = (2S+1) \frac{k_f^3}{6\pi^2} \qquad (1.64)$$

with n denoting the constant particle density in position space of this ground state, related to the Fermi radius k_f. The Fermi sphere of radius k_f in $\boldsymbol{k}$-space separates the occupied orbitals $\langle \boldsymbol{r} | \boldsymbol{k} \rangle$ from the unoccupied ones. The ground state energy is

$$E = (2S+1) \sum_{k \leq k_f} \frac{\hbar^2 k^2}{2m} = \frac{3\hbar^2 k_f^2}{10m} N \qquad (1.65)$$

implying an average energy per particle

$$\varepsilon = \frac{E}{N} = \frac{3\hbar^2 k_f^2}{10m} \qquad (1.66)$$

in the ground state of the homogeneous interaction-free fermion gas.

1.1.4 Heisenberg Representation

The notations (1.27 and 1.58) specify the Schrödinger and momentum representations to be special cases of a more general scheme.

Let $\{|L\rangle\}$ be any given complete orthonormal set of N-particle states labelled by some multi-index L. Any state may then be expanded according to

$$|\Psi\rangle = \sum_L |L\rangle C_L = \sum_L |L\rangle \langle L|\Psi\rangle, \qquad (1.67)$$

and the eigenvalue equation (1.4) takes on the form of a matrix problem:

$$\sum_{L'} [A_{LL'} - a\delta_{LL'}] C_{L'} = 0, \quad A_{LL'} = \langle L|\hat{A}|L'\rangle. \qquad (1.68)$$

1.1 Résumé on N-Particle Quantum States

with an infinite matrix $\|\boldsymbol{A}\|$ and the eigenstate with eigenvalue a represented by a column vector $\boldsymbol{C}$. This is Heisenberg's matrix mechanics.

To be a bit more specific, consider a *fermion* system and let $\{\phi_l(x)\}$ be a complete orthonormal set of single-particle (spin-)orbitals. Let Φ_L, $L = (l_1 \ldots l_N)$ run over the N-particle Slater determinants (1.19) of all possible orbital configurations (again using some linear order of the l-labels). In analogy to (1.59–1.61) one now gets

$$H_{LL'} = \sum_i \left\langle l_{\bar{i}} \left| -\frac{\hbar^2}{2m}\nabla^2 + v \right| l'_i \right\rangle (-1)^{\bar{\mathcal{P}}} \prod_{j(\neq i)} \delta_{l_j l'_j} + $$
$$+ \frac{1}{2} \sum_{i \neq j} \left[\langle l_{\bar{i}} l_{\bar{j}} | w | l'_i l'_j \rangle - \langle l_{\bar{i}} l_{\bar{j}} | w | l'_j l'_i \rangle \right] (-1)^{\bar{\mathcal{P}}} \prod_{k(\neq i,j)} \delta_{l_k l'_k} \quad (1.69)$$

for the Hamiltonian with $i = \bar{\mathcal{P}}\bar{i}$, and $\bar{\mathcal{P}}$ is defined in the same way as in (1.59), particularly again $\bar{\mathcal{P}}$ = identity for $L' = L$. The orbital matrix elements are

$$\langle l | \hat{h} | m \rangle \stackrel{\text{def}}{=} \left\langle l \left| -\frac{\hbar^2}{2m}\nabla^2 + v \right| m \right\rangle =$$
$$= \sum_{ss'} \int d^3r \, \phi_l^*(\boldsymbol{r}, s) \left[-\frac{\hbar^2}{2m}\nabla^2 \delta_{ss'} + v_{ss'}(\boldsymbol{r}) \right] \phi_m(\boldsymbol{r}, s') \quad (1.70)$$

and

$$\langle lm | w | pq \rangle = \sum_{s_1 s'_1 s_2 s'_2} \int d^3r_1 d^3r_2 \, \phi_l^*(\boldsymbol{r}_1, s_1) \phi_m^*(\boldsymbol{r}_2, s_2) \, *$$
$$* \, w_{s_1 s'_1, s_2 s'_2}(\boldsymbol{r}_1, \boldsymbol{r}_2) \, \phi_q(\boldsymbol{r}_2, s'_2) \phi_p(\boldsymbol{r}_1, s'_1). \quad (1.71)$$

Clearly, $\langle lm | w | pq \rangle = \langle ml | w | qp \rangle$.

1.1.5 Hartree-Fock Theory

For an interacting N-fermion system, a single Slater determinant (1.19) can of course not be a solution of the stationary Schrödinger equation. However, one can ask for the best Slater determinant approximating the true N-particle ground state as that one which minimizes the expectation value of the Hamiltonian $\hat{H}$ among Slater determinants. The corresponding minimum value will estimate the true ground state energy from above.

This would, however, in general be a too restrictive search. The point is that in most cases of interest the Hamiltonian $\hat{H}$ does not depend on the spins of the particles: $v_{ss'}(\boldsymbol{r}) = \delta_{ss'}v(\boldsymbol{r})$ and $w_{s_1s_1',s_2s_2'}(\boldsymbol{r}_1, \boldsymbol{r}_2) = \delta_{s_1s_1'}\delta_{s_2s_2'}w(|\boldsymbol{r}_1 - \boldsymbol{r}_2|)$. Consequently, the true ground state has a definite total spin $S^2 = \langle \sum_\alpha^{x,y,z}(\sum_i \hat{\sigma}_{i\alpha})^2 \rangle$, whereas a Slater determinant of spin-orbitals in general does not have a definite total spin; rather such a spin eigenstate can be build as a linear combination of Slater determinants with the same spatial orbitals but different single-particle spin states occupied. Depending on whether the total spin of the ground state is zero or non-zero, the approach is called the closed-shell and open-shell Hartree-Fock method, respectively.

We restrict our considerations to the simpler case of closed shells of spin-half particles and will see in a minute that in this special case a determinant of spin-orbitals would do. In this case, the number N of spin-half particles must be even because otherwise the total spin would again be half-integer and could not be zero. A spin-zero state of two spin-half particles is obtained as the antisymmetric combination of a spin-up and a spin-down state:

$$\langle s_1 s_2 | S = 0 \rangle = \frac{1}{\sqrt{2}} \left(\chi^+(s_1)\chi^-(s_2) - \chi^-(s_1)\chi^+(s_2) \right). \tag{1.72}$$

This is easily seen by successively operating with $\hat{\sigma}_{1\alpha} + \hat{\sigma}_{2\alpha}, \alpha = x, y, z$ (see (1.10, 1.15)) on it, giving a zero result in all cases. Hence, a simple product of $N/2$ spin pairs in states (1.72) provides a normalized N-particle $S = 0$ spin state, which is antisymmetric with respect to particle exchange within the pair and symmetric with respect to exchange of pairs. (It cannot in general be symmetric or antisymmetric with respect to exchange between different pairs.)

The two particles in the spin state (1.72) may occupy the same spatial orbital $\phi(\boldsymbol{r})$, maintaining the antisymmetric character of the pair wavefunction

$$\Phi(x_1 x_2) = \phi(\boldsymbol{r}_1)\phi(\boldsymbol{r}_2)\langle s_1 s_2 | S = 0 \rangle. \tag{1.73}$$

If, for even N, we consider a Slater determinant of spin orbitals, where each spatial orbital is occupied twice with spin up and down, and expand the determinant into a sum over permutations of products, then a permutation within a doubly occupied orbital does not change the spatial part of those terms. For the spin part, all those permutations just combine to a product of $N/2$ spin-zero states (1.72). Now take *any* permuted product appearing in the expansion of the determinant. It has been either part of the just considered combination or not. If not, take again all permutations within pairs of equal spatial orbitals (which all have not been

1.1 Résumé on N-Particle Quantum States

part of the former combination) to get another product of $N/2$ spin-zero states. The total Slater determinant is thus a linear combination of products of spin-zero states, hence it is itself a spin-zero state in this special case. Moreover, as a Slater determinant it has the correct antisymmetry with respect to all particle exchange operations. Therefore, such a single Slater determinant can provide a spin-zero approximant to a closed-shell ground state.

Now, take (for even N) $N/2$ spatial orbitals ϕ_i and build a Slater determinant with spin-orbitals $\phi_{i+}(x_k)$ in the first $N/2$ rows and with spin-orbitals $\phi_{i-}(x_k)$ in the lower $N/2$ rows. Using the Laplace expansion, this Slater determinant may be written as

$$\Phi_{\mathrm{HF}}(x_1 \ldots x_N) = \frac{(N/2)!}{\sqrt{N!}} \sum_{\{k|k'\}} \frac{(-1)^{\{k|k'\}}}{(N/2)!} \det \|\phi_{i+}(x_k)\| \det \|\phi_{i-}(x_{k'})\|, \tag{1.74}$$

where $\{k|k'\}$ means a selection of $N/2$ numbers k among the numbers $1,\ldots,N$, the remaining unselected numbers being denoted by k'. There are $N!/(N/2)!^2$ different selections to be summed up with an appropriately chosen sign for each item. The items of the sum are normalized, and they are orthogonal to each other with respect to their spin dependence, because they differ in the selection of the variables of spin-up particles. Hence there are no crossing matrix elements for any spin independent operator, and its expectation value may be calculated just with one of the terms in the sum of (1.74), all terms giving the same result.

With the help of (1.69–1.71) the expectation value of the Hamiltonian (with s-independent external potential v) in the state (1.74) is easily obtained to be

$$E_{\mathrm{HF}} = \langle \Phi_{\mathrm{HF}} | \hat{H} | \Phi_{\mathrm{HF}} \rangle = 2 \sum_{i=1}^{N/2} \langle i | \hat{h} | i \rangle + 2 \sum_{i,j=1}^{N/2} \langle ij | w | ij \rangle - \sum_{i,j=1}^{N/2} \langle ij | w | ji \rangle. \tag{1.75}$$

The three terms are called in turn one-particle energy, Hartree energy, and exchange energy. Summation over both spin directions for each orbital ϕ_i results in factors 2 for the one-particle term, 4 for the Hartree term, but only 2 for the exchange term because the contained matrix element is only nonzero if both interacting particles have the same spin direction. (Recall that the interaction part of the Hamiltonian comes with a pre-factor $1/2$.) Note that both the Hartree and exchange terms for $i = j$ contain (seemingly erroneously) the self-interaction of a particle in the orbital

ϕ_i with itself. Actually those terms of the Hartree and exchange parts mutually cancel in (1.75) thus not posing any problem.

In order to find the minimum of this type of expression one must vary the orbitals keeping them orthonormal. However, as we already know, the determinant remains unchanged upon an orthogonalization of the orbitals, and hence it suffices to keep the orbitals normalized while varying them. Adding the normalization integral for ϕ_k, multiplied with a Lagrange multiplier $2\varepsilon_k$, to (1.75) and then varying ϕ_k^* leads to the minimum condition

$$\hat{h}\,\phi_k(\boldsymbol{r}) + v_\mathrm{H}(\boldsymbol{r})\,\phi_k(\boldsymbol{r}) + (\hat{v}_\mathrm{X}\,\phi_k)(\boldsymbol{r}) = \phi_k(\boldsymbol{r})\varepsilon_k \tag{1.76}$$

with the Hartree potential

$$v_\mathrm{H}(\boldsymbol{r}) = 2\sum_{j=1}^{N/2} \int d^3r'\, \phi_j^*(\boldsymbol{r}')\, w(|\boldsymbol{r}-\boldsymbol{r}'|)\, \phi_j(\boldsymbol{r}') \tag{1.77}$$

and the exchange potential operator

$$(\hat{v}_\mathrm{X}\,\phi_k)(\boldsymbol{r}) = -\sum_{j=1}^{N/2} \int d^3r'\, \phi_j^*(\boldsymbol{r}')\, w(|\boldsymbol{r}-\boldsymbol{r}'|)\, \phi_k(\boldsymbol{r}')\phi_j(\boldsymbol{r}). \tag{1.78}$$

(ϕ_k^* is varied independently of ϕ_k which is equivalent to independently varying the real and imaginary parts of ϕ_k; the variation then is carried out by using the simple rule $\delta/\delta\phi_k^*(x)\int dx'\,\phi_k^*(x')F(x') = F(x)$ for any expression $F(x)$ independent of $\phi_k^*(x)$.)

The Hartree-Fock equations (1.76) have the form of effective single-particle Schrödinger equations

$$\hat{F}\,\phi_k = \phi_k\,\varepsilon_k, \tag{1.79}$$

where the Fock operator $\hat{F} = -(\hbar/2m)\nabla^2 + \hat{v}_\mathrm{eff}$ consists of the kinetic energy operator and an effective potential operator

$$\hat{v}_\mathrm{eff} = v + v_\mathrm{H} + \hat{v}_\mathrm{X} \tag{1.80}$$

called the mean field or molecular field operator.

For a given set of $N/2$ occupied orbitals ϕ_i the Fock operator $\hat{F}$ as an integral operator is the same for all orbitals. Hence, from (1.79), the Hartree-Fock orbitals may be obtained orthogonal to each other. From (1.76) it then follows that

$$\sum_{i=1}^{N/2} \varepsilon_i = \sum_{i=1}^{N/2} \langle i|\hat{h}|i\rangle + 2\sum_{i,j=1}^{N/2} \langle ij|w|ij\rangle - \sum_{i,j=1}^{N/2} \langle ij|w|ji\rangle. \tag{1.81}$$

Comparison with (1.75) yields

$$E_{\mathrm{HF}} = \sum_{i=1}^{N/2} (\varepsilon_i + \langle i|\hat{h}|i\rangle) = 2\sum_{i=1}^{N/2} \varepsilon_i - \langle \hat{W}\rangle \tag{1.82}$$

for the total Hartree-Fock energy. The sum over all occupied ε_i (including the spin sum) double-counts the interaction energy.

Coming back to the expression (1.75), one can ask for its change, if one removes one particle in the Hartree-Fock orbital ϕ_k (of one spin direction) *while keeping all orbitals ϕ_j un-relaxed*. This change is easily obtained to be $-\langle k|\hat{h}|k\rangle - 2\sum_j \langle kj|w|kj\rangle + \sum_j \langle kj|w|jk\rangle$, which is just $-\varepsilon_k$ as seen from (1.76–1.78). For a given set of occupied ϕ_i, (1.79) yields also unoccupied orbitals as solutions. The change of (1.75), if one additionally occupies one of those latter orbitals ϕ_k, is analogously found to be $+\varepsilon_k$. These results, which may be written as

$$\left(\frac{\partial E_{\mathrm{HF}}}{\partial n_k}\right)_{\phi_j} = \varepsilon_k \tag{1.83}$$

with n_k denoting the occupation number of the Hartree-Fock orbital ϕ_k and the subscript ϕ_j indicating the constancy of the orbitals, goes under the name Koopmans' theorem [Koopmans, 1934]. It guarantees in most cases that the minimum of E_{HF} is obtained if one occupies the orbitals with the lowest ε_i, because removing a particle from ϕ_i and occupying instead a state ϕ_j yields a change of E_{HF} equal to $\varepsilon_j - \varepsilon_i$ plus the orbital relaxation energy, which is usually smaller than $\varepsilon_j - \varepsilon_i$ in closed shell situations.

1.2 The Fock Space

Up to here we considered representations of quantum mechanics with the particle number N of the system fixed. If this number is macroscopically large, it cannot be fixed at a single definite number in experiment. Zero mass bosons as e.g. photons may be emitted or absorbed in systems of any scale. (In a relativistic description any particle may be created or annihilated, possibly together with its antiparticle, in a vacuum region just by applying energy.) From a mere technical point of view, quantum statistics of identical particles is much simpler to formulate with the grand canonical ensemble with varying particle number, than with the canonical one. Hence there are many good reasons to consider quantum dynamics with changes in particle number.

In order to do so, we start with building the Hilbert space of quantum states of this wider frame: the Fock space. The considered up to now Hilbert space of all N-particle states having the appropriate symmetry with respect to particle exchange will be denoted by $\mathcal{H}_N$. In Subsection 1.1.4 an orthonormal basis $\{|L\rangle\}$ of (anti-)symmetrized products of single-particle states out of a given fixed complete and ortho-normalized set $\{\phi_i\}$ of such single-particle states was introduced. *The set $\{\phi_i\}$ with some fixed linear order $(\phi_1, \phi_2, \ldots)$ of the orbitals will play a central role in the present section.* The *normalized* states $|L\rangle$ will alternatively be denoted by

$$|n_1 \ldots n_i \ldots\rangle, \quad \sum_i n_i = N, \tag{1.84}$$

where n_i denotes the *occupation number* of the i-th single-particle orbital in the given state $|L\rangle$. For fermions, $n_i = 0, 1$, for bosons $n_i = 0, 1, 2, \ldots$. Two states (1.84) not coinciding in all occupation numbers n_i are *orthogonal*. $\mathcal{H}_N$ is the complete linear space spanned by the basis vectors (1.84), i.e. the states of $\mathcal{H}_N$ are either linear combinations $\sum |L\rangle C_L$ of states (1.84) (with the sum of the squared absolute values of the coefficients C_L equal to unity) or limits of Cauchy sequences of such linear combinations. A Cauchy sequence is a sequence $\{|\Psi_n\rangle\}$ with $\lim_{m,n\to\infty} \langle \Psi_m - \Psi_n | \Psi_m - \Psi_n \rangle = 0$. The inclusion of all limits of such sequences into $\mathcal{H}_N$ means realizing the *topological* completeness property of the Hilbert space, being extremely important in all considerations of limits. This completeness of the space is not to be confused with the completeness of a basis set $\{\phi_i\}$. The extended Hilbert space $\mathcal{F}$ (Fock space) of all states with the particle number N not fixed is now defined as the completed direct sum of all $\mathcal{H}_N$. It is spanned by all state vectors (1.84) for *all* N with the above given definition of orthogonality retained, and is completed by corresponding Cauchy sequences, just as the real line is obtained from the rational line by completing it with the help of Cauchy sequences of rational numbers. (A mathematical rigorous treatment can be found, e.g. in [Cook, 1953, Berezin, 1965].)

Note that $\mathcal{F}$ now contains not only quantum states which are linear combina-

1.2 The Fock Space

tions with varying n_i so that n_i does not have a definite value in the quantum state (occupation number fluctuations), but also linear combinations with varying N so that now quantum fluctuations of the total particle number are allowed too. For bosonic fields (as e.g. laser light) those quantum fluctuations can become important experimentally even for macroscopic N.

1.2.1 Occupation Number Representation

In order to introduce the possibility of a dynamical change of N, operators must be introduced providing such a change. For bosons those operators are introduced as

$$\hat{b}_i |\ldots n_i \ldots\rangle = |\ldots n_i - 1 \ldots\rangle \sqrt{n_i}, \tag{1.85}$$

$$\hat{b}_i^\dagger |\ldots n_i \ldots\rangle = |\ldots n_i + 1 \ldots\rangle \sqrt{n_i + 1}. \tag{1.86}$$

These operators annihilate and create, respectively, a particle in the orbital ϕ_i and multiply by a factor chosen for the sake of convenience. Particularly, in (1.85) it prevents producing states with negative occupation numbers. (Recall that the n_i are integers; application of $\hat{b}_i$ to a state with $n_i = 0$ gives zero instead of a state with $n_i = -1$.) Considering all possible matrix elements with the basis states (1.84) of $\mathcal{F}$, one easily proves that $\hat{b}$ and $\hat{b}^\dagger$ are Hermitian conjugate to each other. In the same way the key relations

$$\hat{n}_i |\ldots n_i \ldots\rangle \stackrel{\text{def}}{=} \hat{b}_i^\dagger \hat{b}_i |\ldots n_i \ldots\rangle = |\ldots n_i \ldots\rangle n_i, \tag{1.87}$$

and

$$[\hat{b}_i, \hat{b}_j^\dagger] = \delta_{ij}, \quad [\hat{b}_i, \hat{b}_j] = 0 = [\hat{b}_i^\dagger, \hat{b}_j^\dagger] \tag{1.88}$$

are proven, where the brackets in standard manner denote the commutator $[\hat{b}_i, \hat{b}_j^\dagger] \stackrel{\text{def}}{=} [\hat{b}_i, \hat{b}_j^\dagger]_- = \hat{b}_i \hat{b}_j^\dagger - \hat{b}_j^\dagger \hat{b}_i$. The occupation number operator $\hat{n}_i$ is Hermitian and can be used to define the particle number operator

$$\hat{N} = \sum_i \hat{n}_i \tag{1.89}$$

having arbitrarily large but *always finite* expectation values in the basis states (1.84) of the Fock space $\mathcal{F}$. The Fock space itself is the complete hull (in the above described sense) of the linear space spanned by all possible states obtained from the normalized vacuum state

$$|\rangle \stackrel{\text{def}}{=} |0\ldots 0\ldots\rangle, \quad \hat{b}_i|\rangle = 0 \text{ for all } i \tag{1.90}$$

by applying polynomials of the $\hat{b}_i^\dagger$ to it. This situation is expressed by saying that the vacuum state is a cyclic vector of $\mathcal{F}$ with respect to the algebra of the $\hat{b}_i$ and $\hat{b}_i^\dagger$. Obviously, any operator in $\mathcal{F}$, that is any operation transforming vectors of $\mathcal{F}$ linearly into new ones, can be expressed as a power series of operators $\hat{b}_i^\dagger$ and $\hat{b}_i$. This all together means that the Fock space provides an irreducible representation space for the algebra of operators $\hat{b}_i^\dagger$ and $\hat{b}_i$, defined by (1.88).

For fermions, the definition of creation and annihilation operators must have regard for the antisymmetry of the quantum states and for Pauli's exclusion principle following from this antisymmetry. They are defined by

$$\hat{c}_i|\ldots n_i\ldots\rangle = |\ldots n_i - 1\ldots\rangle\, n_i\, (-1)^{\sum_{j<i} n_j}, \tag{1.91}$$

$$\hat{c}_i^\dagger|\ldots n_i\ldots\rangle = |\ldots n_i + 1\ldots\rangle\, (1 - n_i)\, (-1)^{\sum_{j<i} n_j}. \tag{1.92}$$

Again by considering the matrix elements with all possible occupation number eigenstates (1.84), it is easily seen that these operators have all the needed properties, do particularly not create non-fermionic states (that is, states with occupation numbers n_i different from 0 or 1 do not appear: application of $\hat{c}_i$ to a state with $n_i = 0$ gives zero, and application of $\hat{c}_i^\dagger$ to a state with $n_i = 1$ gives zero as well). The $\hat{c}_i$ and $\hat{c}_i^\dagger$ are mutually Hermitian conjugate, obey the key relations

$$\hat{n}_i|\ldots n_i\ldots\rangle \stackrel{\text{def}}{=} \hat{c}_i^\dagger \hat{c}_i|\ldots n_i\ldots\rangle = |\ldots n_i\ldots\rangle\, n_i \tag{1.93}$$

and

$$[\hat{c}_i, \hat{c}_j^\dagger]_+ = \delta_{ij}, \quad [\hat{c}_i, \hat{c}_j]_+ = 0 = [\hat{c}_i^\dagger, \hat{c}_j^\dagger]_+ \tag{1.94}$$

with the anti-commutator $[\hat{c}_i, \hat{c}_j^\dagger]_+ = \hat{c}_i \hat{c}_j^\dagger + \hat{c}_j^\dagger \hat{c}_i$ defined in standard way. Their role in the fermionic Fock space $\mathcal{F}$ is completely analogous to the bosonic case. (The $\hat{c}^\dagger$- and $\hat{c}$-operators of the fermionic case form a normed complete algebra provided with a norm-conserving adjoint operation †, called a c^*-algebra in mathematics. Such a (normed) c^*-algebra can be formed out of the bosonic operators $\hat{b}^\dagger$ and $\hat{b}$, which themselves are not bounded in $\mathcal{F}$, by complex exponentiation. For a comprehensive treatment on the level of mathematical physics see [Bratteli and Robinson, 1987].)

1.2 The Fock Space

As an example, the Hamiltonian (1.69) is expressed in terms of creation and annihilation operators and orbital matrix elements (1.70, 1.71) as

$$\hat{H} = \sum_{ij} \hat{c}_i^\dagger \langle i|\hat{h}|j\rangle \hat{c}_j + \frac{1}{2} \sum_{ijkl} \hat{c}_i^\dagger \hat{c}_j^\dagger \langle ij|w|kl\rangle \hat{c}_l \hat{c}_k \qquad (1.95)$$

Observe the order of operators being important in expressions of that type; this Hamiltonian is indeed equivalent to (1.69) for every N since it does not change the total particle number. The present form is easily verified by considering the matrix element $\langle L|\hat{H}|L'\rangle$ with $|L\rangle$ and $|L'\rangle$ represented in notation (1.84), and comparing the result with (1.69).

Generally, an operator is said to be in *normal order*, if it is arranged in each term so that all creators are left of all annihilators. The result of normal-ordering an operator $\hat{A}$ is indicated by colons, $:\hat{A}:$. It is obtained by just (anti)commuting the factors in all terms of $\hat{A}$. By applying the (anti)commutation rules for the $\hat{b}$- and $\hat{c}$-operators, every operator can be identically transformed into a sum of normal-ordered operators. For instance, $:\hat{b}\hat{b}^\dagger: = \hat{b}^\dagger \hat{b}$, but $\hat{b}\hat{b}^\dagger = \hat{b}^\dagger \hat{b} + 1 = :\hat{b}\hat{b}^\dagger: + 1$. The merit of normal order is that vacuum expectation values vanish for all terms containing at least one creation or annihilation operator.

In order to write down some useful relations holding accordingly in both the bosonic and fermionic cases, we use operator notations $\hat{a}_i$ and $\hat{a}_i^\dagger$ denoting either a bosonic or a fermionic operator. One easily obtains

$$[\hat{n}_i, \hat{a}_i] = -\hat{a}_i, \quad [\hat{n}_i, \hat{a}_i^\dagger] = \hat{a}_i^\dagger \qquad (1.96)$$

with the *commutator* in both the bosonic and fermionic cases.

Sometimes it is useful (or simply hard to be avoided) to use a non-orthogonal basis $\{\phi_i\}$ of single-particle orbitals. The whole apparatus may be generalized to this case by merely generalizing the first relations (1.88) and (1.94) to

$$[\hat{a}_i, \hat{a}_j^\dagger]_\pm = \langle \phi_i|\phi_j\rangle, \qquad (1.97)$$

which generalization of course comprises the previous relations of the orthogonal cases. Even with a non-orthogonal basis $\{\phi_i\}$ the form of the original relations (1.88) and (1.94) may be retained, if one defines the operators $\hat{a}_i$ with respect to the ϕ_i and replaces the operators $\hat{a}_i^\dagger$ by modified creation operators $\hat{a}_i^+$ with respect to a contragredient basis $\{\chi_i\}$, $\langle \phi_i|\chi_j\rangle = \delta_{ij}$. Of course, this way the $\hat{a}_i^+$ are no longer Hermitian conjugate to the $\hat{a}_i$.

1.2.2 Coherent Bosonic States

The states (1.84) are simultaneously eigenstates of all occupation number operators $\hat{n}_i = \hat{a}_i^\dagger \hat{a}_i$, $\hat{a}_i = \hat{b}_i$ or $\hat{a}_i = \hat{c}_i$, as seen from (1.87) and (1.93). Most of those states are not eigenstates of the creation or annihilation operators as is likewise seen from (1.85, 1.86) and (1.91, 1.92). It is easily verified that a bosonic creator $\hat{b}_i^\dagger$ has no ordinary eigenstate at all in the Fock space. (Exercise. Take a general state $\sum |L\rangle C_L$ and let N_L be the total particle number in the component $|L\rangle$. Let N_{L_0} be the minimum number N_L for which C_L is non-zero. It exists since $N_L \geq 0$ and $N_L < \infty$. Apply $\hat{b}_i^\dagger$.)

Remarkably enough, besides the vacuum there exist *simultaneous* eigenstates of all bosonic annihilators $\hat{b}_i$ (the $\hat{b}_i$ commute with one another) in the Fock space [Glauber, 1963]. For reasons which become clear at the end of this subsection, they are called *coherent states*.

Consider the eigenvalue equations (for all i simultaneously)

$$\hat{b}_i |\boldsymbol{b}\rangle = |\boldsymbol{b}\rangle b_i \tag{1.98}$$

and use the occupation number notation

$$|\boldsymbol{b}\rangle = \sum |n_1 \ldots n_i \ldots\rangle C_{n_1 \ldots n_i \ldots}. \tag{1.99}$$

From (1.85) it follows by projection onto one basis state that

$$C_{n_1 \ldots n_i \ldots} \sqrt{n_i} = C_{n_1 \ldots n_i - 1 \ldots} b_i. \tag{1.100}$$

The state $|n_1 \ldots n_i \ldots\rangle$ can be created out of the vacuum state (1.90) as

$$|n_1 \ldots n_i \ldots\rangle = \frac{(\hat{b}_1^\dagger)^{n_1}}{\sqrt{n_1!}} \cdots \frac{(\hat{b}_i^\dagger)^{n_i}}{\sqrt{n_i!}} \cdots |\rangle. \tag{1.101}$$

(Induction of (1.86).) The last two relations yield, again by induction,

$$\begin{aligned} |\boldsymbol{b}\rangle &= \sum_{n_1 \ldots n_i \ldots} \frac{(b_1 \hat{b}_1^\dagger)^{n_1}}{n_1!} \cdots \frac{(b_i \hat{b}_i^\dagger)^{n_i}}{n_i!} \cdots |\rangle = \\ &= \exp\left(\sum b_i \hat{b}_i^\dagger\right) |\rangle. \end{aligned} \tag{1.102}$$

Here and in the following, the sum in the exponent runs over the subscript, i in the present case. The b_i may be arbitrary complex numbers.

1.2 The Fock Space

Hermitian conjugation of equations (1.98) and (1.102) yields

$$\langle\boldsymbol{b}|\hat{b}_i^\dagger = b_i^* \langle\boldsymbol{b}| \qquad (1.103)$$

wherefore $\langle\boldsymbol{b}|$ is called a *left eigenstate* of the $\hat{b}_i^\dagger$, and

$$\langle\boldsymbol{b}| = \langle| \exp\left(\sum \hat{b}_i b_i^*\right). \qquad (1.104)$$

Hence,

$$\langle\boldsymbol{b}|\boldsymbol{b}'\rangle = \langle| \exp\left(\sum \hat{b}_i b_i^*\right) \exp\left(\sum b_j' \hat{b}_j^\dagger\right) |\rangle = \exp\left(\sum b_i^* b_i'\right). \qquad (1.105)$$

After rewriting the two exponentials of sums into products of exponentials and expanding the latter into series, the nonzero vacuum expectation values are those containing factors $\hat{b}_i$ and $\hat{b}_i^\dagger$ in pairs only. Considering (1.88) yields the final result (exercise).

We conclude that (generally non-normalized) coherent states are *all* states (1.102), for which $\sum |b_i|^2$ is finite, the complex numbers b_i may otherwise be completely arbitrary. This may be expressed by saying that the annihilators $\hat{b}_i$ have a complex continuous spectrum. Moreover, by expanding the exponential in (1.102) we find

$$\langle|\boldsymbol{b}\rangle = 1 \qquad (1.106)$$

for all $|\boldsymbol{b}\rangle$.

Applying a creator to a (right) coherent state (1.102) yields

$$\hat{b}_i^\dagger |\boldsymbol{b}\rangle = \hat{b}_i^\dagger \exp\left(\sum b_j \hat{b}_j^\dagger\right) |\rangle = \frac{\partial}{\partial b_i} \exp\left(\sum b_j \hat{b}_j^\dagger\right) |\rangle = \frac{\partial}{\partial b_i} |\boldsymbol{b}\rangle \qquad (1.107)$$

and likewise, from (1.104),

$$\langle\boldsymbol{b}| \hat{b}_i = \frac{\partial}{\partial b_i^*} \langle\boldsymbol{b}|. \qquad (1.108)$$

This can be used to evaluate the following commutator:

$$\left[\hat{b}_i, \int \prod_j \frac{d^2 b_j}{\pi} |\boldsymbol{b}\rangle \exp\left(-\sum b_k^* b_k\right) \langle\boldsymbol{b}|\right] =$$

$$= \int \prod_j \frac{d^2 b_j}{\pi} |\boldsymbol{b}\rangle \exp\left(-\sum b_k^* b_k\right) \left(b_i - \frac{\partial}{\partial b_i^*}\right) \langle\boldsymbol{b}|$$

$$= 0. \qquad (1.109)$$

The measure $d^2z = dxdy$, $z = x+iy$ was introduced in the complex plane. In the second line, (1.98) and (1.108) was used. Instead of using the independent variables Re b_j, Im b_j in the complex b_j-plane, b_j^* and b_j may likewise be considered independent, with $d\text{Re } b_j d\text{Im } b_j = \partial(\text{Re } b_j, \text{Im } b_j)/\partial(b_j^*, b_j)\, db_j^* db_j = db_j^* db_j/2i$, hence $d^2 b_j/\pi = db_j^* db_j/(2\pi i)$. Then, $|\boldsymbol{b}\rangle$ depends on b_j, that is, on the combination Re $b_j + i\text{Im } b_j$ only, and $\langle \boldsymbol{b}|$ depends on b_j^* only. An integration by parts gives a zero result in (1.109). The Hermitian conjugate of this result is

$$\left[\int \prod_j \frac{d^2 b_j}{\pi} |\boldsymbol{b}\rangle \exp\left(-\sum b_k^* b_k\right) \langle \boldsymbol{b}|, \hat{b}_i^\dagger\right] = 0. \tag{1.110}$$

Now, recall that the Fock space is an irreducible representation space of the algebra of the $\hat{b}_i$ and the $\hat{b}_i^\dagger$. That implies, that an operator commuting with all $\hat{b}_i$ and all $\hat{b}_i^\dagger$ must be proportional to the unit operator (Schur's lemma). Hence the above multi-integral must be a c-number. Its value may be obtained by taking the vacuum expectation value. Considering (1.106) one finds that

$$\int \prod_j \frac{d^2 b_j}{\pi} |\boldsymbol{b}\rangle \exp\left(-\sum b_k^* b_k\right) \langle \boldsymbol{b}| = 1 \tag{1.111}$$

is the coherent state representation of the unit operator in the bosonic Fock space. Note, however, that the coherent states $|\boldsymbol{b}\rangle$ are not orthogonal to each other as explicitly seen from (1.105, 1.106).

In analogy to (1.67, 1.68), a coherent state representation for every quantum state $|\Psi\rangle$ and for every operator $\hat{A}$ in the Fock space may be introduced. With the help of the unit operator (1.111),

$$|\Psi\rangle = \int \prod \frac{d^2 b_i}{\pi} |\boldsymbol{b}\rangle \exp\left(-\sum b_k^* b_k\right) \Psi(\boldsymbol{b}^*), \quad \Psi(\boldsymbol{b}^*) \stackrel{\text{def}}{=} \langle \boldsymbol{b}|\Psi\rangle. \tag{1.112}$$

Here, $\Psi(\boldsymbol{b}^*) = \Psi(b_1^* \ldots b_i^* \ldots)$ figures as a 'wavefunction', compare to $|\Psi\rangle = \int d^3 r\, |\boldsymbol{r}\rangle \langle \boldsymbol{r}|\Psi\rangle$. In analogy to (1.27) we find from (1.108, 1.103)

$$\langle \boldsymbol{b}|\hat{b}_i|\Psi\rangle = \frac{\partial}{\partial b_i^*} \Psi(\boldsymbol{b}^*), \quad \langle \boldsymbol{b}|\hat{b}_i^\dagger|\Psi\rangle = b_i^* \Psi(\boldsymbol{b}^*). \tag{1.113}$$

This yields the annihilation and creation operators

$$\hat{b}_i = \frac{\partial}{\partial b_i^*}, \quad \hat{b}_i^\dagger = b_i^* \tag{1.114}$$

1.2 The Fock Space

in coherent state representation. For instance, the Schrödinger equation with the Hamiltonian (1.95) reads in coherent state representation

$$\left(\sum_{ij} b_i^* \langle i|\hat{h}|j\rangle \frac{\partial}{\partial b_j^*} + \frac{1}{2}\sum_{ijkl} b_i^* b_j^* \langle ij|w|kl\rangle \frac{\partial}{\partial b_k^*}\frac{\partial}{\partial b_l^*}\right)\Psi(\boldsymbol{b}^*) = \Psi(\boldsymbol{b}^*)E. \quad (1.115)$$

The matrix elements are the same as in (1.70, 1.71).

On the other hand, for the matrix elements of a *normal-ordered* Fock space operator $:\hat{A}(\hat{b}_i^\dagger, \hat{b}_j):$ with coherent states one simply finds (exercise)

$$\langle \boldsymbol{b}| :\hat{A}(\hat{b}_i^\dagger, \hat{b}_j): |\boldsymbol{b}'\rangle = A(b_i^*, b_j') \exp\left(\sum b_k^* b_k'\right), \quad (1.116)$$

where $A(b_i^*, b_j')$ means that in the algebraic expression of $\hat{A}(\hat{b}_i^\dagger, \hat{b}_j)$ each entry of $\hat{b}_i^\dagger$ is replaced with the complex number b_i^* and each entry of $\hat{b}_j$ with b_j'. For instance, in a coherent state,

$$\langle n_i \rangle = \frac{\langle \boldsymbol{b}|\hat{b}_i^\dagger \hat{b}_i|\boldsymbol{b}\rangle}{\langle \boldsymbol{b}|\boldsymbol{b}\rangle} = |b_i|^2 \quad (1.117)$$

and

$$\frac{\Delta n_i}{\langle n_i \rangle} = \langle n_i \rangle^{-1/2}. \quad (1.118)$$

(Exercise. Transformation to normal order of $\hat{n}_i^2$ is the essential issue.) *In a coherent state, the relative particle number fluctuations vanish for macroscopic mode occupation.*

Consider a single one-dimensional harmonic oscillator with Hamiltonian

$$\hat{H} = \hat{b}^\dagger \hbar\omega \hat{b}, \quad \hat{b}^\dagger = \frac{1}{\sqrt{2\hbar\omega}}(\omega\hat{x} - i\hat{p}), \quad \hat{b} = \frac{1}{\sqrt{2\hbar\omega}}(\omega\hat{x} + i\hat{p}). \quad (1.119)$$

(The mass is put to unity.) For the Schrödinger wavefunction $\psi_b(x) = \langle x|b\rangle$ we have

$$\langle x|b\rangle b = \langle x|\hat{b}|b\rangle = \frac{\langle x|\omega\hat{x} + i\hat{p}|b\rangle}{\sqrt{2\hbar\omega}} = \frac{1}{\sqrt{2\hbar\omega}}\left(\omega x + \hbar\frac{\partial}{\partial x}\right)\langle x|b\rangle, \quad (1.120)$$

i.e.,

$$\frac{\partial}{\partial x}\psi_b(x) = \left(\sqrt{\frac{2\omega}{\hbar}}b - \frac{\omega}{\hbar}x\right)\psi_b(x) \quad (1.121)$$

with the solution

$$\psi_b(x) = C(b) \exp\left(-\left[\sqrt{\frac{\omega}{2\hbar}}\, x - b\right]^2\right). \tag{1.122}$$

C is the integration constant which in general may of course depend on b. This is a minimum uncertainty wave pocket with its center of gravity at $x = \sqrt{2\hbar/\omega}\,\mathrm{Re}\, b$. (For $b = 0$, i.e. the Fock space vacuum, it is just the oscillator ground state; it is normalized to unity with $C(0) = (\omega/\pi\hbar)^{1/4}$. The normalization (1.106) means $\int dx\, \psi_0^*(x)\psi_b(x) = 1$, yielding $C(b) = C(0)\exp(b^2/2)$.) Now, in the Heisenberg picture $\hat{b} \to \hat{b}\exp(-i\omega t)$ and hence, in (1.120), $b \to b\exp(-i\omega t)$: the center of gravity of the wave pocket oscillates with the classical oscillator frequency and an arbitrary amplitude determined by b. The phases of the oscillator quanta are *coherently* related in such a way that the wave pocket moves in the minimum uncertainty shape without decaying. The situation can directly be transferred to photons, where coherent states describe traveling minimum uncertainty light pulses with arbitrary amplitudes. It is this context in which coherent states provide a limiting transition from bosonic quantum states to well localized particle-like classical wave pockets.

1.2.3 Grassmann Numbers

In the fermion case, the only eigenvalue for both creators $\hat{c}_i^\dagger$ and annihilators $\hat{c}_i$ in the *physical* Hilbert space is zero since from (1.94) $\hat{c}_i^{\dagger 2} = 0 = \hat{c}_i^2$. It is, however, useful to introduce a *formal* symmetry between bosons and fermions, which has got the name *super-symmetry* in physical theories, but which is not realized in our world (it is speculated that it could have been spontaneously broken in the world's present state). To this goal, since non-zero complex numbers cannot be eigenvalues of fermionic annihilators, abstract *super-number generators* ζ_i are introduced, which anti-commute

$$\zeta_i\zeta_j + \zeta_j\zeta_i = 0, \text{ i.e. } \zeta_i^2 = 0 \tag{1.123}$$

and hence generate a Grassmann algebra, whence the name Grassmann numbers. (To contrast anti-commuting numbers with commuting numbers, the names *a*-number and *c*-number, respectively, are often used; a product of two *a*-numbers is a *c*-number.) A super-analysis is developed for super-numbers (as well as a super-algebra, super-topology, ...).

1.2 The Fock Space

A general *super-number* is a general holomorphic function (power series) of the super-number generators

$$f(\zeta_i) = z_0 + \sum_i z_i \zeta_i + \sum_{i \neq j} z_{ij} \zeta_i \zeta_j + \cdots, \qquad (1.124)$$

where the z are complex coefficients commuting with every super-number, and each term of the series cannot contain higher then first powers of each variable ζ_i due to the second relation (1.123). The complex number $f_b = z_0$ is said to be the 'body' of f, and the remainder $f_s = f - f_b$ its 'soul'. The soul consists of an a-number part formed by the odd-order terms and a c-number part formed by the even-order terms. Only an f with non-zero body has an inverse obtained as the power series of $(f_b + f_s)^{-1}$ in powers of f_s. For instance the exponential function is

$$\exp \zeta = 1 + \zeta. \qquad (1.125)$$

Thus, with respect to the ζ_i only derivatives of multi-linear functions are needed, and the only peculiarity here is that derivative operators with respect to super-number generators *anti-commute* with a-numbers and with each other:

$$\frac{\partial}{\partial \zeta_j} \zeta_i \zeta_j = -\zeta_i \frac{\partial}{\partial \zeta_j} \zeta_j = -\zeta_i, \quad \frac{\partial}{\partial \zeta_i} \frac{\partial}{\partial \zeta_j} = -\frac{\partial}{\partial \zeta_j} \frac{\partial}{\partial \zeta_i}. \qquad (1.126)$$

Integration $\int d\zeta_i \, f(\zeta_j)$ is defined by the rules that $d\zeta_i$ is again an anti-commuting symbol, an integral of a complex linear combination is equal to the complex linear combination of integrals, the result of integration over $d\zeta_i$ does no longer depend on ζ_i, and, in order to enable integration by parts, *the integral over a derivative is zero*. These rules imply

$$\int d\zeta \, z = 0, \quad \int d\zeta \, z\zeta = z. \qquad (1.127)$$

The first result follows since z is the derivative of $f(\zeta) = z\zeta$, and a constant c-number-factor left open by the above rules in the second result is defined by convention (not uniquely in the literature).

Finally, a conjugation $\zeta_i \to \zeta_i^*$ is introduced which groups the super-number generators into pairs (where ζ_i^* is different from ζ_i, so that $\zeta_i^* \zeta_i$ is non-zero[8]) and

---

[8] Caution, this is a special choice. In super-mathematics the generators of the super-algebra (distinctively denoted θ_k here) usually are defined to be 'real': $\theta_k^* = \theta_k$. Our choice can then be realized by putting $\zeta_k = (\theta_{2k-1} + i\theta_{2k})/\sqrt{2}$, $\zeta_k^* = (\theta_{2k-1} - i\theta_{2k})/\sqrt{2}$.

which obeys the rules

$$(z\zeta)^* = z^*\zeta^*, \quad (\zeta_i\zeta_j)^* = \zeta_j^*\zeta_i^*. \tag{1.128}$$

Observe that conjugate Grassmann number generators just anti-commute as conjugate complex numbers just commute, in contrast to conjugate operators (1.94) and (1.88). A comprehensive introduction into super-mathematics accessible for physicists is given in [DeWitt, 1992].

1.2.4 Coherent Fermionic States

Since coherent fermionic states do not exist in the physical Fock space $\mathcal{F}$, this space first must be generalized: the physical Fock space of fermions is generated by the operators $\hat{c}_i^\dagger$. We attach to each operator $\hat{c}_i$ a Grassmann number generator ζ_i and specify *anti-commutation between Grassmann number generators and $\hat{c}$-operators* as well as *Hermitian conjugation to comprise Grassmann number conjugation in reversed order of all factors in a product*, which also means the order of $\hat{c}$-operators and Grassmann numbers to be reversed in a product. This way a Grassmann algebra is specified and related to the algebra of fermion creators and annihilators. The generalized Fock space $\mathcal{G}$ is now defined as consisting of all linear combinations of vectors of $\mathcal{F}$ with coefficients out of that Grassmann algebra (i.e. of the form (1.124)). The vectors of $\mathcal{F}$ have c-number properties as previously. The general vectors of $\mathcal{G}$ inherit their commutation properties from the coefficients of their expansion into $\mathcal{F}$-vectors.

In analogy to (1.102), except for a fermionic minus sign in the exponent, we consider the state

$$|\zeta\rangle = \exp\left(-\sum \zeta_i \hat{c}_i^\dagger\right) |\rangle = \prod (1 - \zeta_i \hat{c}_i^\dagger) |\rangle. \tag{1.129}$$

Since bilinear terms of anti-commuting quantities commute, the exponential of a sum may simply be written as the product of individual exponentials, which according to (1.125) reduce to linear expressions. Operation with an annihilator

1.2 The Fock Space

on this state yields

$$\begin{aligned}
\hat{c}_i |\zeta\rangle &= \hat{c}_i \prod (1 - \zeta_j \hat{c}_j^\dagger) |\rangle; = \\
&= \left(\prod_{j(\neq i)} (1 - \zeta_j \hat{c}_j^\dagger) \right) (\hat{c}_i + \zeta_i \hat{c}_i \hat{c}_i^\dagger) |\rangle = \\
&= \left(\prod_{j(\neq i)} (1 - \zeta_j \hat{c}_j^\dagger) \right) \zeta_i |\rangle = \\
&= \left(\prod_{j(\neq i)} (1 - \zeta_j \hat{c}_j^\dagger) \right) (1 - \zeta_i \hat{c}_i^\dagger) |\rangle \zeta_i = \\
&= |\zeta\rangle \zeta_i.
\end{aligned} \quad (1.130)$$

In the second line, $\hat{c}_i$ was anti-commuted with ζ_i (while it commutes with all bilinear $\zeta_j \hat{c}_j^\dagger$ for $j \neq i$). This step was anticipated by introducing the fermionic minus sign in the exponent of (1.129). Then, in the third line, $\hat{c}_i |\rangle = 0$ and $\hat{c}_i \hat{c}_i^\dagger |\rangle = |\rangle$ was used. In the fourth line, a 'nutritious' zero was added (observe that $\zeta_i \cdots \zeta_i$ is zero according to the Grassmann rules), whereupon it is seen, that $|\zeta\rangle$ *is a fermionic coherent state*, i.e., an eigenstate of all annihilators $\hat{c}_i$ with Grassmann eigenvalues ζ_i.

Along similar lines one finds

$$\langle\zeta| \hat{c}_i^\dagger = \zeta_i^* \langle\zeta|, \quad \langle\zeta| = \langle| \exp\left(-\sum \hat{c}_i \zeta_i^*\right) \quad (1.131)$$

and

$$\langle\zeta|\zeta'\rangle = \exp\left(\sum \zeta_i^* \zeta_i'\right), \quad \langle|\zeta\rangle = 1 \text{ for all } |\zeta\rangle. \quad (1.132)$$

(Exercise.)

A *physical* fermion state is generated by applying creators to the vacuum: $|n_1 \ldots n_i \ldots\rangle = (\hat{c}_1^\dagger)^{n_1} \cdots (\hat{c}_i^\dagger)^{n_i} \cdots |\rangle$. In view of the eigenvalue equation (1.130) its overlap with a coherent state is

$$\langle n_1 \ldots n_i \ldots |\zeta\rangle = \langle| \cdots (\hat{c}_i)^{n_i} \cdots (\hat{c}_1)^{n_1} |\zeta\rangle = \cdots \zeta_i^{n_i} \cdots \zeta_1^{n_1}. \quad (1.133)$$

Likewise, the matrix element of a *normal-ordered* Fock space operator $:\hat{A}(\hat{c}_i^\dagger, \hat{c}_j):$ with coherent states is

$$\langle\zeta| :\hat{A}(\hat{c}_i^\dagger, \hat{c}_j): |\zeta'\rangle =\, :A(\zeta_i^*, \zeta_j'): \exp\left(\sum \zeta_k^* \zeta_k'\right), \quad (1.134)$$

where the right hand side has a meaning as in (1.116), except that now the order of factors remains essential.

Observing the rules of the last subsection, (1.133) can be used to prove that

$$\int \prod_j d\zeta_j^* d\zeta_j \, |\zeta\rangle \exp\left(-\sum \zeta_k^* \zeta_k\right) \langle \zeta| = 1 \tag{1.135}$$

is the unit operator in the *physical* Fock space $\mathcal{F}$ generated by the states (1.84) for all N (exercise).

Hence, although fermionic coherent states are unphysical (except for the vacuum), *every physical fermionic state can be expanded into coherent states* with the help of (1.135):

$$|\Psi\rangle = \int \prod d\zeta_i^* d\zeta_i \, |\zeta\rangle \exp\left(-\sum \zeta_k^* \zeta_k\right) \Psi(\zeta^*), \quad \Psi(\zeta^*) \stackrel{\text{def}}{=} \langle \zeta|\Psi\rangle. \tag{1.136}$$

Again, $\Psi(\zeta^*) = \Psi(\zeta_1^* \ldots \zeta_i^* \ldots)$ figures as a 'wavefunction' (observe the order of arguments).

From $(1 - \hat{c}_i \zeta_i^*)\hat{c}_i = \hat{c}_i = (\partial/\partial \zeta_i^*)(1 - \hat{c}_i \zeta_i^*)$ (since $\hat{c}_i^2 = 0$, and before applying the derivative it must be anti-commuted with $\hat{c}_i$) as well as a conjugated relation one verifies

$$\langle \zeta| \hat{c}_i = \frac{\partial}{\partial \zeta_i^*} \langle \zeta|, \qquad \hat{c}_i^\dagger |\zeta\rangle = |\zeta\rangle \frac{\overleftarrow{\partial}}{\partial \zeta_i} \;. \tag{1.137}$$

Again,

$$\hat{c}_i = \frac{\partial}{\partial \zeta_i^*}, \quad \hat{c}_i^\dagger = \zeta_i^* \tag{1.138}$$

is the coherent state representation of annihilation and creation operators.

If we finally declare a measure $d\mu_c$ of the coherent configuration space to be

$$d\mu_c \stackrel{\text{def}}{=} \prod_i \frac{d^2 b_i}{\pi} \text{ for bosons}, \quad d\mu_c \stackrel{\text{def}}{=} \prod_i d\zeta_i^* d\zeta_i \text{ for fermions} \tag{1.139}$$

and occasionally introduce a statistics sign factor $\eta = \pm 1$, we obtain a complete super-symmetry[9] in the description of bosons and fermions within the coherent state space $\mathcal{G}$. Recall, however, that our real world does not show this (broken?) super-symmetry: it travels all the time through the subspace $\mathcal{F}$ of $\mathcal{G}$ only. While the subspace $\mathcal{F}$, the Fock space, contains all bosonic coherent states which quantum-mechanically describe real macroscopic fields as e.g. classical electromagnetic fields, it does not contain fermionic coherent states except of the vacuum, there are no macroscopic fermionic fields in nature (not any more?).

For applications of supersymmetry in quantum theory of non-interacting particles in a random potential see [Efetov, 1997].

1.3 Field Quantization

Quantum fields are obtained either by canonical quantization of classical fields via canonical field variables defined with the help of a Lagrange density, or by combining the Schrödinger orbitals and the corresponding creators and annihilators in Fock space. The latter route is often misleadingly called second quantization although there is always only one quantization.

A spatial representation may be introduced in the Fock space by defining field operators

$$\hat{\psi}(x) = \sum_i \phi_i(x) \hat{a}_i, \quad \hat{\psi}^\dagger(x) = \sum_i \phi_i^*(x) \hat{a}_i^\dagger, \tag{1.140}$$

which obey the relations

$$\begin{aligned}[] [\hat{\psi}(x), \hat{\psi}^\dagger(x')]_\pm &= \delta(x-x'), \\ [\hat{\psi}(x), \hat{\psi}(x')]_\pm &= 0 = [\hat{\psi}^\dagger(x), \hat{\psi}^\dagger(x')]_\pm. \end{aligned} \tag{1.141}$$

and provide a spatial particle density operator

$$\hat{n}(x) = \hat{\psi}^\dagger(x)\hat{\psi}(x) \tag{1.142}$$

---

[9]*Formally* this symmetry can be made even more explicit by using the generators θ_k of the footnote on page 43, observing $(d\theta_k)^2 = 0$, and defining $\int d\theta\, z\theta = \sqrt{i\pi} z$ at variance with (1.127). This would imply

$$d\mu_c \stackrel{\text{def}}{=} \prod_k \frac{d\theta_{2k} d\theta_{2k-1}}{\pi} \text{ for fermions.}$$

Since we are not really going to use super-symmetry we did not burden our analysis with formal factors here.

having the properties

$$\langle n(x)\rangle = \sum_{ij} \phi_i^*(x)\langle \hat{a}_i^\dagger \hat{a}_j\rangle \phi_j(x), \qquad \int dx\, \hat{n}(x) = \sum_i \hat{a}_i^\dagger \hat{a}_i. \qquad (1.143)$$

These relations are readily obtained from those of the creation and annihilation operators, and by taking into account the completeness and ortho-normality

$$\sum_i \phi_i(x)\phi_i^*(x') = \delta(x-x'), \qquad \int dx\, \phi_i^*(x)\phi_j(x) = \delta_{ij} \qquad (1.144)$$

of the basis orbitals.

We may introduce a spin density matrix operator

$$\hat{n}_{ss'}(\boldsymbol{r}) = \hat{\psi}^\dagger(\boldsymbol{r}s')\hat{\psi}(\boldsymbol{r}s) \qquad (1.145)$$

in accordance with (1.45) (which is its expectation value), and a vector spin density operator

$$\hat{\boldsymbol{S}}(\boldsymbol{r}) = \frac{\hbar}{2}\,\text{tr}\,\hat{n}(\boldsymbol{r})\hat{\boldsymbol{\sigma}} \qquad (1.146)$$

in accordance with (1.47). Particularly, the spin flip operators are

$$\hat{S}_+(\boldsymbol{r}) = \hbar\hat{\psi}^\dagger(\boldsymbol{r}+)\hat{\psi}(\boldsymbol{r}-), \qquad \hat{S}_-(\boldsymbol{r}) = \hbar\hat{\psi}^\dagger(\boldsymbol{r}-)\hat{\psi}(\boldsymbol{r}+). \qquad (1.147)$$

They rise and lower the z-component of the spin by one quantum.

In terms of field operators, the Hamiltonian (1.26) reads

$$\hat{H} = \int dx\, \hat{\psi}^\dagger(x) \sum_{s'} \left[-\frac{\hbar^2}{2}\delta_{ss'}\nabla^2 + v_{ss'}(\boldsymbol{r})\right] \hat{\psi}(x') +$$

$$+ \frac{1}{2}\int dx_1 dx_2\, \hat{\psi}^\dagger(x_1)\hat{\psi}^\dagger(x_2) \sum_{s_1's_2'} w_{s_1s_1',s_2s_2'}(\boldsymbol{r}_1,\boldsymbol{r}_2)\, \hat{\psi}(x_2')\hat{\psi}(x_1'), \qquad (1.148)$$

which is easily obtained by combining (1.95) with (1.140) and considering (1.75, 1.76). As in (1.27), we use the notation $x' = (\boldsymbol{r}, s')$.

Field-quantized interaction terms contain higher-order than quadratic expressions in the field operators and hence yield operator forms of equations of motion (in Heisenberg picture) which are nonlinear. Note, however, that, contrary to the Fock operator of (1.76), the Hamiltonians (1.95, 1.148) are *linear* operators in the Fock space of states $|\Psi\rangle$ as demanded by the superposition principle of quantum theory. In this respect, the Fock operator rather compares to those operator equations of motion than to a Hamiltonian.

1.3 Field Quantization

1.3.1 Locality

As seen from (1.140), quantum field operators—or in short, quantum fields—are spatial densities of annihilators or creators. Hence, they are operators *potentially making up or changing fields in space* rather then directly representing physical fields being present in the laboratory. The latter appear as properties of quantum states on which quantum field operators may act.

As potential fields, quantum fields must be defined in the whole infinite space-time. However, any actual manipulation on a physical system—which in quantum physics is spoken of as a measuring process—takes place in a *finite* domain of space (and time). Hence, *observables* must be local constructs of quantum fields. For instance, a local orbital $\phi_k(x)$ is annihilated by $\int dx\, \phi_k^*(x)\hat{\psi}(x) = \hat{a}_k$, which immediately follows from the definition (1.140) of $\hat{\psi}(x)$ and the ortho-normality of the orbital set $\{\phi_k\}$. Even this local annihilator is generally not yet an observable, because the process must be accompanied by something else, e.g. taking over the energy and the spin of the annihilated state.

Consider arbitrarily often differentiable functions $f, g, \ldots$ vanishing outside of a bounded closed domain of $\boldsymbol{r}$-space. Define

$$\hat{\psi}(f) \stackrel{\text{def}}{=} \int dx\, f^*(x)\hat{\psi}(x), \quad \hat{\psi}^\dagger(g) \stackrel{\text{def}}{=} \int dx\, \hat{\psi}^\dagger(x)g(x). \tag{1.149}$$

From (1.141),

$$\begin{aligned}[][\hat{\psi}(f), \hat{\psi}^\dagger(g)]_\pm &= \int dx\, dx'\, f^*(x)[\hat{\psi}(x), \hat{\psi}^\dagger(x')]_\pm g(x') = \\ &= \int dx\, f^*(x) g(x) = \\ &\stackrel{\text{def}}{=} (f|g) \end{aligned} \tag{1.150}$$

follows, and

$$[\hat{\psi}(f), \hat{\psi}(g)]_\pm = 0 = [\hat{\psi}^\dagger(f), \hat{\psi}^\dagger(g)]_\pm. \tag{1.151}$$

The differentiability of the functions $f, g, \ldots$ together with their decay for large $|\boldsymbol{r}|$ ensures that independently of (1.140) derivatives of quantum fields can be given a definite meaning through integration by parts of expressions like (1.149). Mathematically this means that quantum fields are operator-valued distributions. This gives differential equations of motion for quantum fields a definite mathematical content.

If the domains of f and g are spatially separated, corresponding manipulations on the system should be independent of each other, which means that the corresponding observables should commute. (In relativistic theory this refers to the case that one domain is outside of the light cone of the other.) This corresponds to $(f|g) = 0$ in that case. However, since fermion fields *anti*commute, observables must necessarily be composed of products of even numbers of fermion fields.

As we see, quantum fields by themselves are generally not observable. Observables are local constructs of quantum fields containing only products of even numbers of fermion fields.

1.3.2 Superselection

In Subsection 1.2.1 the algebra of creators $\hat{c}_i^\dagger$ and annihilators $\hat{c}_i$ was linked to the Fock space in such a way, that the latter is an irreducible representation space of the former algebra. If there is a *finite* number of generators of the algebra (finite set of subscripts i), then it can be shown that all non-trivial irreducible representations[10] are unitarily equivalent to this Fock space representation. (Every matrix representation $(\hat{c}_i^\dagger)_{LL'}$, $(\hat{c}_i)_{LL'}$ with respect to *any* complete orthogonal set $\{|L\rangle\}$—not necessarily linked to single-particle orbital occupations—can be unitarily transformed into (1.85, 1.86) or (1.91, 1.92), respectively.)

Consider as a related example the Pauli matrices (1.10). They have the algebraic properties (1.29) reducing every product of Pauli matrices to a linear combination of 1_2 and the σ_γ. Hence, the whole algebra generated by the Pauli matrices consists of these linear combinations only. One may take (1.29), with 1_2 replaced by a unity of yet unknown dimension, as the defining relations for Pauli matrices and ask for all possible non-trivial irreducible representations. These algebraic relations are equivalent to (1.28) and hence contain the angular momentum commutation relations (in units of $\hbar/2$) and the relations $\hat{\sigma}_\alpha^2 = 1$. Thus, $\hat{\sigma}_z$ may only have eigenvalues ± 1, and the only irreducible representations are two-dimensional.[11] Every non-trivial irreducible representation $\hat{\sigma}'_\alpha$ of (1.29) has the form $\hat{\sigma}'_\alpha = \hat{T}\hat{\sigma}_\alpha \hat{T}^{-1}$, where $\hat{\sigma}_\alpha$ are the ordinary Pauli matrices (1.10), and $\hat{T}$ is any regular complex 2×2 matrix. (It must be a unitary 2×2 matrix, if Hermiticity of $\hat{\sigma}'_\alpha$ is to be maintained.)

---

[10]Recall that for every algebra there is the trivial representation which replaces every element of the algebra by zero.

[11]This situation is not to be confused with the situation regarding the *group* $SU(2)$ or the *Lie algebra* $su(2)$ of *commutators* of the $\hat{\sigma}_\alpha$ instead of simple products, where infinitely many different irreducible representations of all integer dimensions exist.

1.3 Field Quantization

This equivalence of all irreducible representations is lost for quantum fields in an infinite space. As one variant of the standard textbook counterexample we consider a regular *infinite* lattice of spin-half states at lattice points $\boldsymbol{l} = (l_x, l_y, l_z)$. At each lattice point we suppose a spinor state

$$\chi_{\boldsymbol{l}}(s) = \langle s|\chi_{\boldsymbol{l}}\rangle = \begin{pmatrix} \chi_{\boldsymbol{l}+} \\ \chi_{\boldsymbol{l}-} \end{pmatrix}, \quad |\chi_{\boldsymbol{l}+}|^2 + |\chi_{\boldsymbol{l}-}|^2 = 1, \tag{1.152}$$

on which the Pauli matrices $\hat{\sigma}_{\boldsymbol{l}\alpha}$, $\alpha = x, y, z$ of (1.10) act. For simplicity we restrict the spinor components to real values,

$$\chi_{\boldsymbol{l}}(s) = \begin{pmatrix} \cos\phi_{\boldsymbol{l}} \\ \sin\phi_{\boldsymbol{l}} \end{pmatrix} = [\cos\phi_{\boldsymbol{l}} - i\sin\phi_{\boldsymbol{l}}\,\hat{\sigma}_{\boldsymbol{l}y}]\begin{pmatrix} 1 \\ 0 \end{pmatrix}. \tag{1.153}$$

The expectation values of the spin components in this state are

$$\begin{aligned}
\langle\sigma_{\boldsymbol{l}x}\rangle &= \langle\chi_{\boldsymbol{l}}|\hat{\sigma}_{\boldsymbol{l}x}|\chi_{\boldsymbol{l}}\rangle = 2\cos\phi_{\boldsymbol{l}}\sin\phi_{\boldsymbol{l}} = \sin 2\phi_{\boldsymbol{l}}, \\
\langle\sigma_{\boldsymbol{l}y}\rangle &= 0, \\
\langle\sigma_{\boldsymbol{l}z}\rangle &= \cos 2\phi_{\boldsymbol{l}}.
\end{aligned} \tag{1.154}$$

(Complex spinor components would allow for a non-zero y-component of the spin expectation value, which does not lead to new aspects in the following consideration.)

Now, imagine a magnetic moment of one magneton attached with each spin and think of an applied external magnetic field, directed in the negative z-direction and homogeneous in the whole infinite space. Then the energetic ground state of the spin lattice is that with

$$\chi_{\boldsymbol{l}}^{(0)}(s) = \begin{pmatrix} 1 \\ 0 \end{pmatrix} \text{ for all } \boldsymbol{l}. \tag{1.155}$$

It corresponds to a homogeneous magnetization density

$$\boldsymbol{m} = (0, 0, 1) \text{ magneton/site.} \tag{1.156}$$

Excitations may be obtained by flipping some of the spins, formally by applying some of the operators

$$\hat{\sigma}_{\boldsymbol{l}}^{\dagger} = \frac{1}{2}(\hat{\sigma}_{\boldsymbol{l}x} - i\hat{\sigma}_{\boldsymbol{l}y}) \tag{1.157}$$

with the properties
$$[\hat{\sigma}_l, \hat{\sigma}_l^\dagger]_+ = 1, \quad [\hat{\sigma}_l, \hat{\sigma}_l]_+ = 0 = [\hat{\sigma}_l^\dagger, \hat{\sigma}_l^\dagger]_+,$$
$$[\hat{\sigma}_l, \hat{\sigma}_l^\dagger]_- = \hat{\sigma}_{lz}, \quad [\hat{\sigma}_l, \hat{\sigma}_{l'}]_- = 0 \text{ for } l \neq l'. \tag{1.158}$$

Hence, we call the homogeneous ground state (1.155) the 'vacuum' $|\rangle$ and the spin-flip operator $\hat{\sigma}_l^\dagger$ a creator of a local spin excitation

$$\hat{\sigma}_l^\dagger |\rangle = |\ldots n_l = 1 \ldots \rangle. \tag{1.159}$$

The corresponding excitation energy is one quantum of Zeeman energy.

The Fock space $\mathcal{F}$ is by definition the norm-completed linear hull of all states containing an arbitrary but *finite* number of local spin excitations. Hence, every state of the Fock space $\mathcal{F}$ is arbitrarily close in the norm-topology to a state with an arbitrarily large but *finite* deviation of the total magnetic moment from its 'vacuum value'. Occupation number eigenstates $|\ldots n_l \ldots\rangle$ containing occupation numbers of excitations $n_l = 1$ at a finite number of sites l and zeros otherwise form a complete basis in this Fock space, which distinguishes all $\hat{\sigma}_l$ and $\hat{\sigma}_l^\dagger$, and these operators provide transitions between all basis vectors. The Fock space $\mathcal{F}$ forms an irreducible representation space for the algebra generated by (1.158).

Consider now again a homogeneous state, in which all sites have equal spin states, but this time

$$\chi_l^{(\phi)}(s) = \begin{pmatrix} \cos \phi \\ \sin \phi \end{pmatrix} \text{ for all } l \tag{1.160}$$

in a representation in which the local operators (1.158) have the previous matrix expressions. Denote this state by $|\rangle_\phi$. Because of $\langle \chi_l^{(0)} | \chi_l^{(\phi)} \rangle = \cos \phi$, one finds

$$\langle |\rangle_\phi = (\cos \phi)^\infty = 0 \text{ for } \phi \neq 0, \pi. \tag{1.161}$$

From (1.154) it follows that $|\rangle_\phi$ corresponds to a homogeneous magnetization density

$$\boldsymbol{m} = (\sin 2\phi, 0, \cos 2\phi) \text{ magneton/site}. \tag{1.162}$$

Excitations above the state $|\rangle_\phi$ may this time formally be obtained by applying the algebra (1.158) to $|\rangle_\phi$. According to (1.158) and (1.10), a spin flip by 180° from the state (1.160) into the state

$$\begin{pmatrix} \cos(\phi + \pi/2) \\ \sin(\phi + \pi/2) \end{pmatrix} = \begin{pmatrix} -\sin \phi \\ \cos \phi \end{pmatrix} \tag{1.163}$$

1.3 Field Quantization

at site l is obtained by applying $\hat{\sigma}'^\dagger_{\phi l} = \cos^2\phi\, \hat{\sigma}'^\dagger_l - \sin^2\phi\, \hat{\sigma}'_l - \sin 2\phi\, \hat{\sigma}'_{lz}$. If, in analogy to (1.159), we denote this state by $|\ldots n_l = 1\ldots\rangle_\phi$, we may again build a complete basis of states with the same occupation numbers as previously, but this time denoting spin flips by π from the state $|\rangle_\phi$. We obtain again a Fock space, which we denote by $\mathcal{F}_\phi$. We have distinguished Pauli matrices acting in the Fock space $\mathcal{F}_\phi$ by a dash, $\hat{\sigma}'_l : \mathcal{F}_\phi \to \mathcal{F}_\phi$, although they are the same matrices (1.10) as those acting in the Fock space $\mathcal{F}$: $\hat{\sigma}_l : \mathcal{F} \to \mathcal{F}$. The algebraic relations between the operators $\{\hat{\sigma}'^\dagger_{\phi l}, \hat{\sigma}'_{\phi l}\}$ are the same as for the operators $\{\hat{\sigma}^\dagger_l, \hat{\sigma}_l\}$. Both sets of operators form two different representations, of the algebra given by the relations (1.158).

In order to examine the relations between $\mathcal{F}$ and $\mathcal{F}_\phi$, consider the scalar product between *any* of the basis states of $\mathcal{F}$ with *any* of the basis states of $\mathcal{F}_\phi$. For all but a finite number of sites, the scalar products of the site spinor states give $\cos\phi$. Hence, as for the two vacua in (1.161), all these scalar products are zero for $\phi \neq 0, \pi$. By norm-continuity of the scalar product, this means that every state of $\mathcal{F}_\phi$ is orthogonal to every state of $\mathcal{F}$:

$$\mathcal{F}_\phi \perp \mathcal{F} \text{ for } \phi \neq 0, \pi. \tag{1.164}$$

Moreover, the algebra generated by (1.158) does not provide transitions from $\mathcal{F}$ to $\mathcal{F}_\phi$.

We introduce a bijective mapping $\hat{R}_\phi : \mathcal{F} \to \mathcal{F}_\phi$ from $\mathcal{F}$ onto $\mathcal{F}_\phi$, which consists in a rotation of the spin vector direction by an angle 2ϕ around the y-axis at *every* lattice site l. Then, obviously $\hat{\sigma}'_{\phi l} = \hat{R}_\phi \hat{\sigma}_l \hat{R}_\phi^{-1}$, and the representations $\{\hat{\sigma}'^\dagger_{\phi l}, \hat{\sigma}'_{\phi l}\}$ and $\{\hat{\sigma}^\dagger_l, \hat{\sigma}_l\}$ of (1.158) are equivalent.

Besides the operators $\{\hat{\sigma}'^\dagger_{\phi l}, \hat{\sigma}'_{\phi l}\}$, the ordinary Pauli matrices $\{\hat{\sigma}'^\dagger_l, \hat{\sigma}'_l\}$ act also in the Fock space $\mathcal{F}_\phi$, and represent also the relations (1.158). This irreducible representation of (1.158) in the space $\mathcal{F}_\phi$ is, however, not equivalent to the previous ones. Since $|\rangle$ and $|\rangle_\phi$ are the only homogeneous states in the spaces $\mathcal{F}$ and $\mathcal{F}_\phi$, respectively, a bijective linear mapping $\hat{T}$ of $\mathcal{F}_\phi$ onto $\mathcal{F}$, which would mediate an equivalence between both representations, must necessarily map $|\rangle_\phi$ onto $|\rangle$: $\hat{T}|\rangle_\phi = |\rangle$ (as $\hat{R}_\phi^{-1}$ above does it). Moreover, $\hat{T}\hat{\sigma}'_l|\rangle_\phi = \hat{\sigma}_l\hat{T}|\rangle_\phi = \hat{\sigma}_l|\rangle = 0$ for all l. Since $\hat{\sigma}'_l|\rangle_\phi \neq 0$, $\hat{T}$ cannot be bijective since an operator which maps a nonzero vector to zero cannot have an inverse. Hence, the mapping $\hat{T}$ with the demanded properties does not exist. Of course, the non-equivalence of the representations follows already from the difference of the magnetization densities between *all* states of both spaces with respect to the direction of those spin flippers.

One could, of course, form the direct sum $\mathcal{F} \oplus \mathcal{F}_\phi$ of both representation spaces and extend the algebra of quantum fields by including additional operators which would provide transitions between $\mathcal{F}$ and $\mathcal{F}_\phi$. While such direct sums of representation spaces play an important role for representing mixed states, in particular thermodynamic states (see next chapter), there is no physical need to introduce operators which make transitions between sectors $\mathcal{F}$ and $\mathcal{F}_\phi$, since those additional operators would not correspond to physical observables. Every measuring process is local in some space region, and there is no possibility to superimpose state vectors of $\mathcal{F}$ with state vectors of $\mathcal{F}_\phi$, i.e. to form a *coherent* superposition with a definite phase relation of the amplitudes in the whole infinite real space. We have already seen that causality prevents a single fermion creator from being observable (because it does not commute with observables in remote space regions); hence, states differing by an odd number of fermions cannot either coherently be superimposed.

The total Hilbert space of physical states decomposes into sectors. The quantum-mechanical superposition principle holds within sectors only. Coherent superpositions of states of different sectors cannot be realized in nature; this is the content of superselection rules. The corresponding sectors of the Hilbert space of all physical states are called superselection sectors. Inequivalent irreducible representations of the algebras of local quantum fields such as (1.149) form different superselection sectors. They may be further subdivided into smaller superselection sectors, for instance in those holding odd and even numbers of fermions, respectively.

2 Macroscopic Quantum Systems

This chapter introduces into the peculiarities of a quantum theoretical treatment of actually infinite systems. The reader is supposed to have been through some introductory course of Statistical Physics and to be familiar with the notions of canonical and grand canonical ensemble. In the first section the thermodynamic limit is considered, which leads to a rigorous classification of microscopic and macroscopic observables and to the important feature of the existence of inequivalent irreducible representations of the operator algebra corresponding to thermodynamic phases. Then, the notion of quantum state is extended to be a certain class of linear functions on the operator algebra. This includes mixed states having been given by density matrices in the quantum mechanics of finite systems. The fundamental Gelfand-Naimark-Segal construction of a Hilbert space from a quantum state is considered. The third section deals with thermodynamic states chosen as canonical density matrix states in the finite case. Their characterization by the Kubo-Martin-Schwinger condition extends to states of actually infinite systems. The last section analyses the unfitness of stationary states for a description of macrosystems and introduces the notion of quasi-stationary excitations instead.

A more detailed treatment of the material of the first three sections may be found in the very recommendable monograph [Sewell, 1986]. A complete mathematical treatise on the theory of normed operator algebras is given in [Bratteli and Robinson, 1987].

2.1 Thermodynamic Limit

In particle physics, one particle or a few particles are considered in an empty homogeneous surroundings described by the Poincaré symmetry. Therefore, one demands that the vacuum must be invariant with respect to the Poincaré group, implying a homogeneous infinite space as a matter of idealization.

In thermodynamics one is interested in the bulk properties of macroscopically homogeneous phases, which do not depend on the presence of grain boundaries. As a matter of idealization, homogeneous states in an infinite volume with properties per unit volume are to be considered as models of thermodynamic phases. Experimentally, a phase grows out of some local nucleation process and spreads into a macroscopic region of space. As we have already seen in the previous chapter, on the level of quantum description, local processes are linked to local observables described by an algebra of local quantum field operators created by the entities (1.149). Besides those local observables, the homogeneously extended state, which corresponds in many respects to the vacuum of particle theory, is characterized by homogeneous *densities*, as for instance particle densities n_α of various kinds α (electrons, various nuclei—or, on a different level of description, phonons—, superfluid condensates, ...), magnetization densities, polarization densities, homogeneous field strengths and so on. The corresponding density operators are expressed typically as

$$\hat{n}_\alpha(\boldsymbol{r}) = \hat{\psi}^\dagger(\boldsymbol{r})\hat{q}_\alpha\hat{\psi}(\boldsymbol{r}) \tag{2.1}$$

through field operators and general 'charge operators' $\hat{q}_\alpha$ which project multi-component field operators onto the relevant charge sector. (We suppress here subscripts for the components of field operators.) If, for instance, the field operators have a spinor structure (like spin-orbitals (1.17)) and $\hat{q}_\alpha = \hat{\sigma}_\alpha$, then $\hat{n}_\alpha(\boldsymbol{r})$ gives the αth Cartesian component of the spin density.

The spatial integral over a density operator gives the operator of the corresponding total 'charge' Q_α in the integration volume of space:

$$\hat{Q}_\alpha = \int_V d^3r\, \hat{n}_\alpha(\boldsymbol{r}), \qquad Q_\alpha = \langle\hat{Q}_\alpha\rangle. \tag{2.2}$$

The above mentioned idealization consists in the thermodynamic limit

$$V \to \infty, \quad Q_\alpha \to \infty, \quad n_\alpha = \frac{Q_\alpha}{V} = \text{constant}. \tag{2.3}$$

For the sake of simplicity of notation, a spatial domain and its volume are here and in the following denoted by the same letter V. We may be tempted to introduce *global* operators

$$\hat{n}_\alpha = \lim_{V \to \infty} \frac{1}{V} \int_V d^3r\, \hat{n}_\alpha(\boldsymbol{r}), \qquad \langle\hat{n}_\alpha\rangle = n_\alpha. \tag{2.4}$$

We shall see in a minute that we don't really need them.

Consider *any* operator $\hat{A}(f)$, build up from expressions of the type (1.149). Then there is some finite spatial domain V_f outside of which f vanishes. Therefore, if $\boldsymbol{r} \notin V_f$, then every field operator $\hat{B}(\boldsymbol{r})$ *of an observable* built from the $\hat{\psi}(\boldsymbol{r})$, $\hat{\psi}^\dagger(\boldsymbol{r})$,

2.1 The Thermodynamic Limit

but containing only products of even numbers of fermion operators, commutes with $\hat{A}(f)$. Hence, in particular,

$$[\hat{n}_\alpha(\boldsymbol{r}), \hat{A}(f)]_- = 0 \text{ if } \boldsymbol{r} \notin V_f. \tag{2.5}$$

The integral

$$\int_V d^3r\, [\hat{n}_\alpha(\boldsymbol{r}), \hat{A}(f)]_- \tag{2.6}$$

is bounded by its finite value for $V = V_f < \infty$. Therefore,

$$[\hat{n}_\alpha, \hat{A}(f)]_- = \lim_{V\to\infty} \frac{1}{V} \int_V d^3r\, [\hat{n}_\alpha(\boldsymbol{r}), \hat{A}(f)]_- = 0 \tag{2.7}$$

for all n_α and all $\hat{A}(f)$. *The global operators commute with all local observables and hence are proportional to the identity operator in every irreducible representation space of the algebra of local observables* (Schur's lemma).

The implications of this situation deserve to be considered a bit more in detail, because the corresponding argumentation has had no relevance in quantum physics of a finite system. Consider an irreducible representation of the algebra of local operators in a Hilbert space $\mathcal{H}_{n_\alpha}$, in which the value of some global density is n_α. (As an example take $\hat{n}_\alpha(\boldsymbol{r}_l) = \hat{\sigma}_{lz}$, $n_\alpha = 1$, $\mathcal{H}_{n_\alpha} = \mathcal{F}$ of Subsection 1.3.2.) The Hilbert space is supposed to be linearly generated by applying arbitrary *local* operators to a certain cyclic vector $|\Psi_0\rangle$, which has the property $\hat{n}_\alpha(\boldsymbol{r})|\Psi_0\rangle = |\Psi_0\rangle n_\alpha$ for *all* $\boldsymbol{r}$. Hence, every state vector $|\Psi\rangle \in \mathcal{H}_{n_\alpha}$ is arbitrarily norm-close to a state vector $|\Psi_{V'}\rangle$ obtained from $|\Psi_0\rangle$ by applying an operator $\hat{R}^\dagger_{V'}$ which commutes with $\hat{n}_\alpha(\boldsymbol{r})$ for points $\boldsymbol{r}$ outside of the *finite* volume V' according to (2.5). Of course, for $\boldsymbol{r} \in V'$, the vector $|\Psi_{V'}\rangle$ need not be an eigenstate of the density operator $\hat{n}_\alpha(\boldsymbol{r})$, that is, generally $\hat{n}_\alpha(\boldsymbol{r})|\Psi_{V'}\rangle \not\propto |\Psi_{V'}\rangle$. However,

$$\begin{aligned}\hat{n}_\alpha(\boldsymbol{r})|\Psi_{V'}\rangle &= \hat{n}_\alpha(\boldsymbol{r})\hat{R}^\dagger_{V'}|\Psi_0\rangle = \hat{R}^\dagger_{V'}\hat{n}_\alpha(\boldsymbol{r})|\Psi_0\rangle = \hat{R}^\dagger_{V'}|\Psi_0\rangle n_\alpha = \\ &= |\Psi_{V'}\rangle n_\alpha \text{ for all } \boldsymbol{r} \notin V'.\end{aligned} \tag{2.8}$$

Now, take a large volume V and subdivide it into N disjunct parts, $V = \sum V_i$. Consider

$$\frac{1}{V}\int_V d^3r\, \hat{n}_\alpha(\boldsymbol{r})|\Psi_{V'}\rangle = \sum_{i=1}^N \frac{1}{V}\int_{V_i} d^3r\, \hat{n}_\alpha(\boldsymbol{r})|\Psi_{V'}\rangle. \tag{2.9}$$

Keep the volume of the domains V_i fixed and let $N \to \infty$, i.e. $V \to \infty$. Then, all but finitely many of the items on the right-hand side give $(V_i/V)|\Psi_{V'}\rangle n_\alpha$. Hence, the result is

$$\hat{n}_\alpha |\Psi\rangle = |\Psi\rangle n_\alpha \text{ for all } |\Psi\rangle \in \mathcal{H}_{n_\alpha}. \qquad (2.10)$$

To be precise, we proved the latter result for all $|\Psi_{V'}\rangle \in \mathcal{H}_{n_\alpha}$, which, however, are norm-dense in the above mentioned sense in $\mathcal{H}_{n_\alpha}$. By making additionally the physically plausible assumption that $\mathcal{H}_{n_\alpha}$ is *locally finite*, which means that the probability to find an infinite number of particles in a finite volume is always zero, the result (2.10) can be extended to all $\mathcal{H}_{n_\alpha}$ by continuity arguments. (In particular, the previously considered Fock space was locally finite, since every state vector $|\Psi\rangle \in \mathcal{F}$ was representable as a Cauchy sequence $|\Psi\rangle = \sum_i |\Psi_i\rangle c_i$ of vectors $|\Psi_i\rangle$ holding a finite number N_i of particles. As a Cauchy sequence, $\lim_{N_i \to \infty} |c_i|^2 = 0$.)

The result (2.10) implies not only

$$[\hat{n}_\alpha, \hat{n}_\beta]_- = 0 \qquad (2.11)$$

for all global densities, but also

$$\Delta n_\alpha = \left(\langle\Psi|\hat{n}_\alpha^2|\Psi\rangle - \langle\Psi|\hat{n}_\alpha|\Psi\rangle^2\right)^{1/2} = 0 \qquad (2.12)$$

for all $|\Psi\rangle \in \mathcal{H}_{n_\alpha}$.

Consequently, in the thermodynamic limit the observables split into two distinct classes: (i) local (microscopic) observables, as previously expressed by local field operators $\hat{\psi}^\dagger(f)$, $\hat{\psi}(f)$ or in particular $\hat{a}_i^\dagger = \hat{\psi}^\dagger(\phi_i)$, $\hat{a}_i = \hat{\psi}(\phi_i)$ for local orbitals ϕ_i, the local observables forming a certain operator algebra, and (ii) homogeneous, that is *spatially averaged* densities n_α of extensive (macroscopic) observables as for instance the average particle density, the average magnetization density, and so on. *Formally* the latter may also be expressed by means of the operator algebra as a non-local limit, but they commute with all local operators and hence have *the same* constant values n_α with zero dispersion ($\Delta n_\alpha = 0$) in all *quantum* states of an *irreducible* representation space of the operator algebra. Their values characterize a *thermodynamic* state of matter in the microcanonical sense. The microscopic quantum dynamics in that thermodynamic state is described with the help of the local operators and the corresponding quantum equations of motion (either, in the Heisenberg picture, for the operators themselves or, in the Schrödinger picture, for the states created with those operators). As was illustrated on a simple example in Subsection 1.3.2, a change of the macroscopic observables is connected with a transition from one irreducible representation of the algebra of local operators to another, inequivalent one. Naturally, this transition will in general also change the microdynamics; we will give a number of examples in subsequent chapters.

Note that, after this discussion, we do *not* include global operators $\hat{n}_\alpha$ in our operator algebra generated by (1.149–1.151), which in the following for short is simply called the operator algebra.

2.2 Pure and Mixed Quantum States

Next, it is necessary to extend our notion of state. Up to here we considered quantum states to be represented by Hilbert space vectors, normalized to unity, and observables to be represented by local operators of an operator algebra in that Hilbert space. The expectation value of the result of measuring an observable A in the state $|\Psi\rangle$ is given by (1.2). It follows immediately that, if 1 is the identity operator, A and B are any two observables, and α and β are complex numbers,

$$\langle(\alpha A + \beta B)\rangle = \alpha\langle A\rangle + \beta\langle B\rangle, \qquad \langle A^\dagger A\rangle \geq 0, \qquad \langle 1\rangle = 1. \tag{2.13}$$

The non-negativity of $\langle A^\dagger A\rangle$ is a direct consequence of the non-negativity of the Hilbert-space norm:

$$\langle A^\dagger A\rangle = \langle\Psi|\hat{A}^\dagger\hat{A}|\Psi\rangle = \langle\Psi'|\Psi'\rangle \geq 0 \text{ with } |\Psi'\rangle = \hat{A}|\Psi\rangle. \tag{2.14}$$

If we write

$$\langle A\rangle = \rho(A), \tag{2.15}$$

we see that a quantum state is a *linear, positive* and *normalized* function ρ on the operator algebra (with complex function values). Linearity, positivity and normalization are in turn expressed by the three relations (2.13).

Given any two linear, positive and normalized functions ρ_1 and ρ_2, it is immediately clear that an affine-linear combination of them with non-negative coefficients, that is, a *convex combination*

$$\rho = c\rho_1 + (1-c)\rho_2, \quad 0 \leq c \leq 1, \tag{2.16}$$

is again a linear, positive and normalized function. Hence, those functions form a convex set.

In quantum mechanics of *finite systems*, a convex combination of orthonormal quantum states $\rho_i(A) = \langle\Psi_i|\hat{A}|\Psi_i\rangle$, that is,

$$\rho(A) = \sum_i p_i \langle\Psi_i|\hat{A}|\Psi_i\rangle = \operatorname{tr}(\hat{\rho}\hat{A}),$$
$$\hat{\rho} = \sum_i |\Psi_i\rangle p_i \langle\Psi_i|, \quad p_i \geq 0, \quad \sum_i p_i = 1, \qquad (2.17)$$

is called a mixed state, given by the density matrix $\hat{\rho}$, which itself is an element of the operator algebra. The mixed states are contrasted with pure states, given by a state vector $|\Psi\rangle$, or alternatively by a density matrix $\hat{\rho}_\Psi = |\Psi\rangle\langle\Psi|$, in which one of the probabilities p_i of (2.17) is equal to unity and all the others are zero.

Because of the orthonormality of the states used, $\langle\Psi_i|\Psi_j\rangle = \delta_{ij}$, we have for every density matrix

$$\hat{\rho}^2 = \sum_i |\Psi_i\rangle p_i^2 \langle\Psi_i| \leq \hat{\rho}, \qquad (2.18)$$

where we mean by the inequality that $\hat{\rho} - \hat{\rho}^2$ provides a positive function $(\rho - \rho^2)(A^\dagger A) = \sum_i (p_i - p_i^2)\langle\Psi_i|\hat{A}^\dagger\hat{A}|\Psi_i\rangle \geq 0$, since $0 \leq p_i \leq 1$ implies $(p_i - p_i^2) \geq 0$, and $\langle\Psi_i|\hat{A}^\dagger\hat{A}|\Psi_i\rangle \geq 0$ as obtained in (2.14). We see also from this consideration, that

$$\hat{\rho}^2 = \hat{\rho} \text{ if and only if } \hat{\rho} = |\Psi\rangle\langle\Psi| \qquad (2.19)$$

for some $|\Psi\rangle$, so that in an expression of the type (2.18) one of the p_i is unity and all the others are zero. Otherwise, ρ^2 is not a state: it is not normalized ($\operatorname{tr}\hat{\rho}^2 < 1$). Suppose now, that $\hat{\rho} = c\hat{\rho}_1 + (1-c)\hat{\rho}_2$, $0 < c < 1$, and that $\hat{\rho}_i$ are pure states, $\hat{\rho}_i^2 = \hat{\rho}_i$, $\operatorname{tr}\hat{\rho}_i = 1$. We have $\hat{\rho}_1\hat{\rho}_2 + \hat{\rho}_2\hat{\rho}_1 = |\Psi_1\rangle\langle\Psi_1|\Psi_2\rangle\langle\Psi_2| + |\Psi_2\rangle\langle\Psi_2|\Psi_1\rangle\langle\Psi_1|$ and hence

$$\operatorname{tr}\hat{\rho}^2 = c^2 + 2|\langle\Psi_1|\Psi_2\rangle|^2 c(1-c) + (1-c)^2 = 1,$$
$$\text{if and only if } |\langle\Psi_1|\Psi_2\rangle| = 1. \qquad (2.20)$$

As $\operatorname{tr}\hat{\rho}^2 = 1$ distinguishes pure states, it follows that a pure state cannot be decomposed into a convex combination of two *different* pure states. Mixed states, on the other hand, decompose by definition into convex combinations of pure states.

A point of a convex set $\mathcal{S}$ in a linear space, which cannot be decomposed into a convex combination of *different* points of that same set $\mathcal{S}$, is said to be an extremal

2.2 Pure and Mixed Quantum States

point of $\mathcal{S}$. Hence, in our context, pure states are precisely the extremal elements of the set of all states.[12]

We are now inclined to call *every* linear, positive and normalized function $\rho(A)$ a state. Since those functions form a convex set, we would distinguish its extremal elements and expect that they correspond precisely to pure states, and the rest to mixed ones. To this end we would have to show that for finite systems every linear, positive and normalized function $\rho(A)$ corresponds to some density matrix $\hat{\rho}$ in our representation space.

As a first step in approaching this problem we demand that all operators of our algebra are bounded (and hence form a c^*-algebra). By physical approximation arguments, that is not a severe modification (see e.g. [Sewell, 1986]). Then we obtain a surprising answer obtained by Gelfand and Naimark, and independently by Segal (GNS construction): [Gelfand and Naimark, 1943, Segal, 1947]

> To *every* linear, positive and normalized function $\rho(A)$ on the operator algebra (c^*-algebra) there exists a Hilbert space $\mathcal{H}_\rho$, a representation of the algebra by linear operators $\hat{A}$ in $\mathcal{H}_\rho$, and a cyclic vector $|\Psi_\rho\rangle \in \mathcal{H}_\rho$, so that $\mathcal{H}_\rho$ is linearly generated by applying the operators $\hat{A}$ to $|\Psi_\rho\rangle$, and
>
> $$\rho(A) = \langle \Psi_\rho | \hat{A} | \Psi_\rho \rangle \text{ for all } \hat{A}. \tag{2.21}$$
>
> This representation is uniquely defined by ρ up to unitary equivalence.[13]

The situation is enlightened by two further results:

i. The representation of the operator algebra, given by the GNS construction, is irreducible if and only if ρ is an extremal element.

---

[12]If x_i are N linear independent points of some linear space, then $x = \sum_i p_i x_i$, $p_i \geq 0$, $\sum_i p_i = 1$ runs through the points of the simplex with corners x_i, in the $(N-1)$-dimensional hyperplane spanned by the points x_i. A simplex is convex, and the so called barycentric decomposition $x = \sum_i p_i x_i$ of its points x is unique, i.e. the numbers p_i are uniquely determined by x. The corners of a simplex form its extremal points. An N-dimensional ball is also convex, and the points of its surface are extremal. However, there are many different barycentric decompositions of the inner points of the ball into extremal points. Caution: The definition of an infinite dimensional simplex is a more subtle issue, and the convex set of all states on the operator algebra is generally not a simplex. Hence, the question of unique decomposition of a general quantum state into pure states is again a subtle issue.

[13]The essential steps of the GNS construction are: (i) to postulate vectors $|\Psi_\rho\rangle$ and $|\Psi_A\rangle = \hat{A}|\Psi_\rho\rangle$ for every $\hat{A}$, (ii) to define the scalar product to be $\langle \Psi_A | \Psi_B \rangle = \rho(A^\dagger B)$, (iii) to identify $|\Psi_A\rangle$ with $|\Psi_{A'}\rangle$ whenever $\rho((A-A')^\dagger(A-A')) = 0$, and (iv) topologically to complete the space.

ii. For a finite quantum system there is up to unitary equivalence only one nontrivial irreducible representation of the operator algebra.

Thus, for a finite quantum system, extremal functions ρ correspond to vector states in the irreducible representation space $\mathcal{H}$ (which is uniquely defined), and the remainder functions ρ correspond to vector states in a reducible representation space of the operator algebra. Every reducible representation space is a direct sum (possibly in some generalized sense[14]) of irreducible representation spaces, in the finite situation where there is essentially only one unique irreducible representation space $\mathcal{H}$, a direct sum of several (possibly infinitely many) copies of $\mathcal{H}$. If in this reducible representation operators and vectors

$$(\hat{A}) = \begin{pmatrix} \hat{A} & 0 & \cdots \\ 0 & \hat{A} & \cdots \\ \vdots & \vdots & \ddots \end{pmatrix}, \quad (|\Psi\rangle) = \begin{vmatrix} \sqrt{p_1}|\Psi_1\rangle \\ \sqrt{p_2}|\Psi_2\rangle \\ \vdots \end{vmatrix} \quad (2.22)$$

are given, then $\langle A \rangle = \sum_i p_i \langle \Psi_i | \hat{A} | \Psi_i \rangle$ as in (2.17): A (normalized) vector state in a reducible representation space means the same as a mixed state in our previous description using only the irreducible representation space $\mathcal{H}$.

Hence, from now on *we will call every linear, positive, normalized (and locally finite in the case of an infinite system) function ρ on the operator algebra a state.* Note that the density matrix $\hat{\rho}$ of (2.17) providing the mixed state ρ was an operator in a single irreducible representation space $\mathcal{H}$. The direct sum of copies of $\mathcal{H}$ was only needed above to cast this state into a vector state.

Two new aspects appear in the thermodynamic limit:

a. There are plenty of inequivalent irreducible representations of the operator algebra (by far not all of which are relevant in physics).

b. There is not any more a one-to-one correspondence between states ρ as linear, positive, normalized and locally finite functions on the operator algebra and density matrices $\hat{\rho}$ as elements of that algebra.

However, given a state ρ and a *bounded* region V of space, $\rho(A)$ is linear, positive and normalized for the set of all operators localized in V (in the sense of (1.149); for every V the weak closure of this set contains the identity operator: take a sequence

---

[14]Generalization can mean that the numbers p_i in (2.22) are to be replaced by some Radon measure, and the sum is replaced by a corresponding integral.

2.2 Pure and Mixed Quantum States

$f_i(x)$ that converges towards $\chi_V(\boldsymbol{r})$, the characteristic function of the region V). Hence there is a density matrix $\hat{\rho}_V$ localized in V (in the same sense) and thus belonging to the operator algebra, for which

$$\rho(A) = \text{tr}\,(\hat{\rho}_V \hat{A}) \text{ for all } \hat{A} \text{ localized in } V. \qquad (2.23)$$

In general, a state ρ corresponds now to a family $\{\hat{\rho}_V\}$ of local density matrices.[15] Since we have many non-equivalent irreducible representations of the operator algebra, the local density matrices of the family *need not lead to a direct sum of copies of a single irreducible representation in the thermodynamic limit*.

Particular quantum states of the macroscopic system are the Fock-space vectors, considered in Subsection 1.3.2. They have the property of short-range correlations. If $\hat{A}(f)$ and $\hat{B}(g)$ are two local *observables* located in V_f and V_g, respectively, then, in the Fock space,

$$\langle A(f) B(g) \rangle - \langle A(f) \rangle \langle B(g) \rangle \to 0 \text{ if } \text{dist}(V_f, V_g) \to \infty. \qquad (2.24)$$

The distance $\text{dist}(V_f, V_g)$ between the domains V_f and V_g is defined to be the lower bound of all distances between points of V_f and V_g, respectively. Recall that $[\hat{A}(f), \hat{B}(g)]_- = 0$, if $\text{dist}(V_f, V_g) > 0$: *observables* always *commute*, since they contain even products of fermion operators only.

The relation (2.24) can be demonstrated as follows. Every state vector of $\mathcal{F}$ can be arbitrarily norm-close approximated by a normalized vector $|\Psi\rangle$ which is locally created out of the normalized vacuum $|\rangle$, and there is an element $\hat{R}^\dagger$ of the local operator algebra, which may be expressed by field operators (1.149) located in some finite volume V_R, with

$$|\Psi\rangle = \hat{R}^\dagger |\rangle,\ \langle |\hat{R}\hat{R}^\dagger|\rangle = 1,\ [\hat{A}(f), \hat{R}]_- = 0 \text{ if } f(\boldsymbol{r}) = 0 \text{ for } \boldsymbol{r} \in V_R. \qquad (2.25)$$

Now, if $\text{dist}(V_f, V_g)$ becomes larger than the largest diameter of V_R, at least either V_f or V_g does not intersect any more V_R. Let V_f not intersect V_R. Then

$$\langle A(f) B(g) \rangle = \langle |\hat{R}\hat{A}(f)\hat{B}(g)\hat{R}^\dagger|\rangle = \langle |\hat{A}(f)\hat{R}\hat{B}(g)\hat{R}^\dagger|\rangle.$$

Transform $\hat{A}(f)$ into normal-ordered terms,

$$\hat{A}(f) = a + \hat{R}_A + \hat{R}_A^\dagger + \sum_i :\hat{A}_i(f):,$$

$$\hat{A}(f)|\rangle = |\rangle a + \hat{R}_A^\dagger|\rangle, \quad \langle|\hat{A}(f) = a\langle| + \langle|\hat{R}_A. \qquad (2.26)$$

---
[15]This family is in fact a net: For every pair $\hat{\rho}_{V_1}, \hat{\rho}_{V_2}$ there is a $\hat{\rho}_V$, $V \supseteq V_1 \cup V_2$ in the family with $\hat{\rho}_{V_i} \leq \hat{\rho}_V$.

Here, a is a c-number, $\hat{R}_A$ is the part containing only annihilation operators so that $\hat{R}_A|\rangle = 0$, and, like $\hat{A}(f)$ itself, all terms commute with $\hat{R}$ and with $\hat{B}(g)$ in the considered limit. We finally have

$$\langle A(f)\rangle = \langle|\hat{R}\hat{A}(f)\hat{R}^\dagger|\rangle = \langle|\hat{A}(f)\hat{R}\hat{R}^\dagger|\rangle = a\langle|\hat{R}\hat{R}^\dagger|\rangle + \langle|\hat{R}\hat{R}^\dagger\hat{R}_A|\rangle = a$$

and

$$\begin{aligned}\langle A(f)B(g)\rangle &= a\langle|\hat{R}\hat{B}(g)\hat{R}^\dagger|\rangle + \langle|\hat{R}\hat{B}(g)\hat{R}^\dagger\hat{R}_A|\rangle = a\langle B(g)\rangle = \\ &= \langle A(f)\rangle\langle B(g)\rangle\end{aligned}$$

which we wanted to demonstrate.

Observe that, if we take a mixed state $\hat{\rho} = |\Psi_0\rangle c\langle\Psi_0| + |\Psi_\phi\rangle(1-c)\langle\Psi_\phi|$ with $|\Psi_0\rangle \in \mathcal{F}$, $|\Psi_\phi\rangle \in \mathcal{F}_\phi$ of the example of Subsection 1.3.2, then the normal-order transformation of $\hat{A}(f)$ yields different constant terms $a_0 \neq a_\phi$ in both Fock spaces because the corresponding annihilation operators $\hat{\sigma}_l$ and $\hat{\sigma}_{\phi l}$ are different. Then it is easily seen that in general (2.24) cannot hold any more (exercise).

The crux here is the property (2.10) with the consequence $\Delta n_\alpha = 0$ and, more generally, $\langle n_\alpha n_\beta\rangle = n_\alpha n_\beta$. If the representation space is not irreducible, these properties are not guaranteed. We conclude that irreducible representations correspond to pure thermodynamic phases, for instance well apart from phase transition points or phase coexistence regions.

2.3 Thermodynamic States

We start again our considerations with *finite quantum systems*. A thermodynamic state is understood as a state of an ensemble of identical copies of the quantum system, so that the (pure) quantum state $|\Psi_i\rangle$ appears with probability p_i in the ensemble. Formally, this leads to the same construct as for a mixed quantum state: ensemble averaging and quantum averaging are treated on the same footing in quantum statistics and cannot be disentangled from one another; thermodynamic states belong to the general set of states introduced in the last section.

As is well known (see any textbook on statistical physics), in this context a thermodynamic equilibrium state at a non-zero temperature T is described by a density matrix, which minimizes the free energy functional[16]

$$F(\hat{\rho}) = \operatorname{tr}(\hat{\rho}\hat{H} + T\hat{\rho}\ln\hat{\rho}), \tag{2.27}$$

---

[16]Here and in the following, the Boltzmann constant is put to $k_B = 1$, i.e., temperatures are measured in units of energy.

2.3 Thermodynamic States

where $\hat{H}$ is the Hamiltonian of the system. Introducing the stationary states of the system contained in a volume V,

$$\hat{H}(V)|\Psi_i\rangle = |\Psi_i\rangle E_i(V), \qquad \langle\Psi_i|\Psi_j\rangle = \delta_{ij}, \tag{2.28}$$

the result of varying (2.27) is the canonical density matrix or statistical operator

$$\hat{\rho}_{T,V} = \frac{1}{Z(T,V)} \sum_i |\Psi_i\rangle e^{-E_i(V)/T} \langle\Psi_i|, \tag{2.29}$$

where the partition function $Z(T,V)$ follows from the normalization condition $\operatorname{tr}\hat{\rho} = 1$ for every density matrix, to be

$$Z(T,V) = \sum_i e^{-E_i(V)/T}. \tag{2.30}$$

Formally, (2.29, 2.30) may be written as

$$\begin{aligned}\hat{\rho}_{T,V} &= \exp\left(-\frac{1}{T}[\hat{H}(V) - F(T,V)]\right), \\ F(T,V) &= -T\ln Z(T,V) = -T\ln \operatorname{tr} \exp(-\hat{H}(V)/T)\end{aligned} \tag{2.31}$$

with the equilibrium free energy $F(T,V)$ as a function of temperature and volume being the minimum value of the functional (2.27).

If the ground state $|\Psi_0\rangle$ of $\hat{H}(V)$ is non-degenerate, then $\hat{\rho}_{T=0,V}$ gives the pure state $|\Psi_0\rangle$, and $F(T \to 0, V) = E_0(V)$.

For the following discussion we introduce the inverse temperature $\beta = T^{-1}$ and, as a complex continuation of time in the Heisenberg picture of operators,

$$\hat{A}(i\hbar\beta) = e^{-\beta\hat{H}} \hat{A} e^{\beta\hat{H}}, \tag{2.32}$$

which is the formal solution of (1.8) for $t = i\hbar\beta$.

Obviously, the linear, positive and normalized function $\rho_{T,V}$, which describes this equilibrium state, obeys the so-called Kubo-Martin-Schwinger condition (KMS condition) [Kubo, 1957, Martin and Schwinger, 1959]

$$\rho_{T,V}(AB) = \rho_{T,V}(BA(i\hbar\beta)) \text{ for all operators } \hat{A}, \hat{B}, \tag{2.33}$$

which is obtained with the representation (2.17):

$$\rho_{T,V}(AB) = \operatorname{tr}(\hat{\rho}_{T,V}\hat{A}\hat{B}) = \operatorname{tr}(\hat{B}\hat{\rho}_{T,V}\hat{A}) = \operatorname{tr}(\hat{B}Z^{-1}e^{-\beta\hat{H}}\hat{A}) =$$

$$= \operatorname{tr}\left(\hat{B}\hat{A}(i\hbar\beta)Z^{-1}e^{-\beta\hat{H}}\right).$$

Except for the definitions of $\hat{\rho}_{T,V}$ and $\hat{A}(i\hbar\beta)$, only the cyclicity of the trace of a product, $\operatorname{tr}(\hat{A}\hat{B}\hat{C}) = \operatorname{tr}(\hat{C}\hat{A}\hat{B})$, was used.

The importance of the KMS condition comes from the fact, that the inverse statement is also true: If a state ρ obeys the KMS condition *for all operators of the operator algebra*, then it corresponds to the canonical density matrix (2.29). This is not difficult to see. Assume that ρ is KMS and that $\hat{\rho}$ is the density matrix giving ρ. Then, for *all* operators $\hat{A}$, $\hat{B}$,

$$\operatorname{tr}\left(\hat{A}\hat{B}\hat{\rho} - \hat{B}e^{-\beta\hat{H}}\hat{A}e^{\beta\hat{H}}\hat{\rho}\right) = \operatorname{tr}\left(\hat{A}(\hat{B}\hat{\rho} - e^{\beta\hat{H}}\hat{\rho}\hat{B}e^{-\beta\hat{H}})\right) = 0$$

and hence for *all* operators $\hat{B}$

$$\hat{B}\hat{\rho}e^{\beta\hat{H}} = e^{\beta\hat{H}}\hat{\rho}\hat{B}.$$

Putting $\hat{B} = 1$ we find $\hat{\rho}\exp(\beta\hat{H}) = \exp(\beta\hat{H})\hat{\rho}$, and hence for all $\hat{B}$, $[\hat{B}, \hat{\rho}\exp(\beta\hat{H})]_- = 0$. An operator commuting with all operators of the algebra is proportional to the identity operator in every irreducible representation space, i.e.,

$$\hat{\rho} \sim e^{-\beta\hat{H}} \text{ and hence } \hat{\rho} = \hat{\rho}_{T,V}.$$

The last step just considers the normalization condition of density matrices. Summarizing, for a finite quantum system the thermodynamic stability condition on $F(\hat{\rho})$ to be minimum and the KMS condition are equivalent. (To be precise, this conclusion makes additionally use of the convexity of $F(\hat{\rho})$.)

In the thermodynamic limit, the (spatial) average of extensive thermodynamic quantities as the (inner) energy, the free energy, the entropy, and so on transform into macroscopic observables taking on constant values over each irreducible representation space, and a mixed state might now mix inequivalent representation spaces characterized by different sets of values of macroscopic observables. The KMS condition continues to be equivalent to local stability, that is, giving a minimum of $F(\hat{\rho}_V)$ by means of local relaxation processes in *every* finite volume V (with $\hat{\rho}_V$ as introduced in (2.23)).

Besides local stability, one can now impose a global stability condition by demanding that the thermodynamic limit $f(\rho) = \lim(1/V)F(\hat{\rho}_V)$ attains its absolute minimum, not only with respect to changes of local observables, but also with respect to macroscopic observables. A large class of model systems with short-range interaction cannot be in a locally stable state without being globally stable. There have been found models with long-range interaction (see for instance [Sewell, 1986]), for which locally stable states exist, which hence are KMS, but not globally stable. It is natural to classify those states as *thermodynamically* metastable. This classification distinguishes them from merely microscopically metastable states, as for instance tunneling states. While tunneling states have a lifetime which tends to a value independent of volume for large volumes, the lifetime of thermodynamically metastable states increases unboundedly in the thermodynamic limit. Then, the KMS condition characterizes precisely the thermodynamically stable and metastable states.

Recall, that the KMS condition (2.33) was expressed for states ρ and not merely for density matrices $\hat{\rho}$. (The importance of this condition for infinite systems was first considered in [Haag et al., 1967].)

2.4 Stationary States

Traditionally, the quantum physics of condensed matter has been interpreted in many treatises in the language of stationary states in analogy to quantum mechanics of atomic systems. As is, however, well known, the basic ingredients of condensed matter theory are quasi-stationary states. This means not simply a slight modification, but leads to far-reaching consequences.

Recall that the main task of experimental and theoretical investigations of atomic systems containing a few degrees of freedom is to find the stationary states Ψ_i. From the point of view of theory this means to find the eigenstates of the Hamiltonian $\hat{H}$, corresponding to eigenvalues E_i or inversely to reconstruct the Hamiltonian from given (experimentally determined) energy eigenvalues E_i and/or matrix elements of physical quantities (observables A). From the experimental point of view the energies E_i and transition matrix elements $(\Psi_i|\hat{A}|\Psi_j)$ of the various observables A are to be measured. The energy eigenvalues E_i and the stationary states Ψ_i (from which all matrix elements $(\Psi_i|\hat{A}|\Psi_j)$ may be calculated) completely determine the behavior of the system. Particularly, the properties of the thermodynamic equilibrium of such a system in a statistical ensemble are expressed in terms of the partition function (2.30) and its derivatives. We consider here the ensemble as a physically real one, for instance a dilute gas of nearly noninteracting atoms or molecules.

It must be emphasized, that, although the interaction of the systems of the ensemble is neglected in (2.30), writing down this formula presupposes the existence of some (arbitrarily weak) interaction. Otherwise the ensemble cannot approach the thermal equilibrium described by this formula. Besides, the stationarity of the states of an atomic system of course implies an approximation: the spontaneous emission of photons (through the coupling of the atom to vacuum fluctuations of the electromagnetic field) allows any atomic system to pass over from an excited 'stationary' level to a lower one, preferably into the ground state. Both these circumstances do not impair the 'central role' of stationary states in quantum mechanics of atomic systems, since both the interaction of the systems (leading to virial corrections to (2.30)) and the intrinsic lifetime due to radiation transitions are calculated in terms of stationary states.

The principal change of this picture when going over to consider macroscopic quantum systems, i.e. systems consisting of a macroscopic number $N \sim 10^{23}$ of *intensely interacting* particles, is just connected with the fact that the stationary states lose their central role. We now have to face a situation where each member of a statistical ensemble contains this macroscopic number of degrees of freedom. Consider as a simple example a very hard material at very low temperature. The situation now described would clearly be even worse for a soft material at elevated temperature. Let the Debye temperature be $\Theta_D = 1000K$, and let the temperature of the sample be $T = 1K$. Let the sample consist of $0.3 \cdot 10^{23}$ atoms, i.e. 10^{23} vibrational degrees of freedom of which at $T/\Theta_D = 10^{-3}$ a fraction of 10^{-9}, that is 10^{14} degrees are active, the rest is 'frozen'. The total vibrational energy of the sample at this temperature T is $E \approx 10^{14}T = 10^{11}\Theta_D$. This energy can be realized with various occupations among all the 10^{23} modes corresponding to different microscopic states. We estimate the number of possibilities from below. One possibility would be to occupy a single mode of energy $\hbar\omega_i \approx \Theta_D/2$ with $2 \cdot 10^{11}$ quanta. Another possibility would be to occupy $2 \cdot 10^{11}$ different modes with $\hbar\omega_i \approx \Theta_D/2$ (there are roughly 10^{23} such modes) with one quantum each. The total number of possibilities is in a crude estimate the number of different selections of 10^{11} modes out of 10^{23}, with multiple selection of the same mode allowed. This number is

$$\binom{10^{23} + 10^{11}}{10^{11}} \approx \binom{10^{23}}{10^{11}} = \frac{10^{23}!}{10^{11}!(10^{23} - 10^{11})!} \approx \frac{(10^{23})^{(10^{11})}}{(10^{11})^{(10^{11})}} \approx$$
$$\approx (10^{12})^{(10^{11})} \approx 10^{(10^{12})}. \qquad (2.34)$$

We used the estimates $M! \approx M^M$ and $N!/(N - M)! \approx N^M$. Furthermore, the last estimate in (2.34) realizes that 'with human means one cannot do any harm to an

2.4 Stationary States

exponent as large as 10^{11}.' Hence, the number of stationary quantum states with energies below $E = 10^{14}$K is estimated by (2.34). This estimate is not affected by anharmonicity, that is, by mode-mode interaction. The *density of stationary quantum states* is then roughly estimated to be

$$D(E) \approx \frac{10^{(10^{12})}}{10^{14}\text{K}} \approx 10^{(10^{12})}\text{K}^{-1}. \tag{2.35}$$

This is a colossal number in any commonly used system of units.

The distance in energy, $\Delta E \sim 10^{-10^{12}}$K is so exorbitantly small, that there is not the least hope to experimentally single out one *definite* stationary state. Needless to say that such a measurement would obviously take an infinite time (estimated from $\Delta E \cdot t \geq \hbar$; with the value of ΔE from above this time t is large compared to the age of our universe, by a still unimaginably huge factor). The evident impossibility to calculate such a state (as the solution of an eigenvalue problem for a partial differential equation in 10^{23} variables) is a common textbook statement.[17] In fact do these considerations already apply to cases of more than thousand degrees of freedom at $T \approx \Theta_D$. In this case the exponent 10^{12} would have to be replaced by 10^3:

$$\binom{10^3 + 10^3}{10^3} = \frac{(2 \cdot 10^3)!}{(10^3!)^2} \approx 2^{2 \cdot 10^3} \approx 10^{10^3},$$

and the above text would still fully apply. The necessary preparation time t would still be about 10^{1000} years while our universe exists for about 10^{10} years. Hence, what follows partly regards large molecules or heavy nuclei too.

Not only the energy cannot be fixed with the desired precision to select a special stationary state, also the particle number itself of a macroscopic system is never exactly fixed due to the interaction with the surrounding medium (or with the measuring device) allowing for the exchange of particles. Theoretically, a system with varying particle number is described by the apparatus of occupation number representation or field quantization, employing the grand canonical ensemble. Though, as regards statistical physics, the use of the grand canonical ensemble is merely dictated by technical convenience and, although fixing the particle number in the system, the canonical ensemble could be used just as well, there is yet a more fundamental reason for applying the apparatus of field quantization. As we

---

[17] Of course, these considerations do not concern the *formal* use of stationary states in developing the theory, for instance in deriving the spectral representation of the Green's function.

have seen in Section 1.3, it provides a natural approach to quantum field theory, where reactions of particles are considered too (their creation, annihilation, and transformation into each other). In the following we shall see that there is no principal difference between the creation of electron-positron pairs by photons in the vacuum and the creation of electron-hole pairs by phonons in a semiconductor.

Experimentally, macroscopic systems are investigated either by scattering, absorption or emission of test particles, the energy, momentum, and polarization of which being measured before and after their interaction with the system, or by direct measurements of thermodynamic and/or kinetic quantities, in particular the various (generalized) susceptibilities. The complex microscopic motion in the macrosystem yields a diffuse contribution to the measured quantities (cross sections, susceptibilities) manifesting in a smooth dependence on both the parameters of the system (e.g. temperature, magnetic field,...) and on the parameters of the operation (e.g. frequency of external fields). If, however, a *quasi-stationary* mode is excited in the system, that is not a really stationary but long-living (on the microscopic time scale) state in which the excitation is concentrated mainly in a certain one-particle or collective degree of freedom (with which, for instance, the test particle interacts), this will manifest in a peak in the cross section or in a well defined behavior of the susceptibilities.

The existence of such quasi-stationary collective modes (degrees of freedom), allowing for a description in terms of single-particle characteristics (having approximately additive energies and momenta), is a typical feature of macroscopic systems. Just this feature makes a microscopic theory of the dynamical properties of a condensed system possible: the description of the 'complex' motion of the macrosystem by means of 'simple' (one-particle) motions. Note that the one-particle state in condensed matter by nature is quasi-stationary, since the procedure of its determination has an approximate character in itself. The corresponding modes interact, and this interaction (which, contrary to the case of the above considered dilute gas, cannot be made arbitrarily weak by experimental conditions) limits the lifetime of the one-particle state.

The properties of isolated particles (atoms or molecules) are studied by investigating a weakly non-ideal gas of atoms or molecules. Conditions (on the density of the gas) may be found, that with any wanted accuracy the total energy of the gas is the sum of energies of the gas particles

$$E_{\text{tot}} = \sum_n \left(E_i^{(n)} + E_{\text{kin}}^{(n)} \right) \tag{2.36}$$

2.4 Stationary States

where $E_{\text{kin}}^{(n)}$ is the kinetic energy of the translational motion of the n-th particle, and $E_i^{(n)}$ is its internal energy. The sum is over all particles. The partition function in this case is

$$\begin{aligned} Z_{\text{tot}}(T,V) &= \prod_n \left[Z^{(n)}(T,V) Z_{\text{kin}}^{(n)}(T,V) \right] = \\ &= [Z(T,V) Z_{\text{kin}}(T,V)]^N \end{aligned} \quad (2.37)$$

where $Z(T,V)$ is defined by (2.30). A main goal in setting up a theory of a condensed system is to find such a set of quasi-stationary excitations which yield, albeit approximately, additive energies. *Densities of states appearing in the context of condensed matter physics refer, in contrast to (2.35), without exception to the approximately additive energies of those elementary excitations.*

In this way, what is seen in experiment and, hence, what quantum theory of macrosystems should be concerned with are not the stationary but the quasi-stationary states being excited by bombarding the system with various test particles in experiment and being created by properly chosen field operators in theory.

3 Quasi-Stationary Excitations

Up to here we essentially considered a quantum state and observables at some given instant of time. Now we start to consider the *dynamics* of the macroscopic quantum system. Time dependences may be treated in different pictures. The algebraic properties of operators, particularly the canonical (anti)commutation relations (1.141) or (1.150, 1.151) of field operators, are valid at a given instant of time, or are generally valid in the Schrödinger picture. They do not hold for time-dependent operators of the Heisenberg picture or—if perturbation theory is considered—of the interaction picture, corresponding to *different* times, which are simultaneously used in expressions describing the propagation of excitations in the quantum system as in particular Green's functions.

Time-dependent Green's functions for zero temperature are introduced in Section 3.2, and the explicit expressions in the trivial case of non-interacting particles are given in Section 3.3. In the two subsequent sections, the connection of poles in the complex frequency dependence of Green's functions with *elementary* quasi-stationary excitations in the quantum system is considered, which includes susceptibilities as the Green's functions for collective excitations. In the last section, Green's functions at non-zero temperature are shortly considered.

3.1 Pictures

In order to describe the dynamics of a quantum system, according to (1.7) or (1.8) a Hamiltonian must be given. Since we are going to use a quantum field representation, the Hamiltonian will have the general form of (1.148), with the only difference that from now on we will generally use a grand canonical Hamiltonian

$$\hat{H}_\mu \stackrel{\text{def}}{=} \hat{H} - \mu \hat{N}. \tag{3.1}$$

Here, μ is the chemical potential, and $\hat{N}$ is the total particle number operator (1.89), or

$$\hat{N} = \int dx \, \hat{\psi}^\dagger(x) \hat{\psi}(x) \tag{3.2}$$

3.1 Pictures

In cases of several types of particles which cannot be transformed into each other (charge superselection sectors), a term $-\mu\hat{N}$ must be introduced for each type in (3.1). With our convention (1.12), $\hat{N}$ of (3.2) is possibly summed over spin orientations. Particles with different spin projections have of course the same chemical potential, if the spin may dynamically flip.

Recall, that at a given temperature T the chemical potential $\mu(T)$ has the meaning of the energy an additional particle must have when added to a thermal equilibrium N-particle state, *in order to obtain a thermal equilibrium $(N + 1)$-particle state.* Hence, given T and $\mu = \mu(T)$, the Hamiltonian $\hat{H}_\mu$ measures the excitation energies from thermal equilibrium, with changes in particle number allowed. (If the particle number would be fixed, the term $-\mu N$ would merely mean an unimportant additive gauge constant for the energy.)

Note further, that for a finite system the Hamiltonian is an extensive observable of the type (2.2), which yields a global density $e - \mu n = \lim_{V \to \infty}(\hat{H}_\mu/V)$ of the grand canonical energy in the thermodynamic limit. Here, however, it is used (without dividing by V) as an *extensive operator form* providing the dynamics of the system. Since we do not divide it by the volume V, it does not commute with the field operators.

In the Schrödinger picture, the dynamics is obtained as a time dependence of the states while the operators of observables are defined independent of time. In the case of a pure state, its time-dependence follows from (1.7) to be

$$|\Psi(t)\rangle = e^{-i\hat{H}_\mu t/\hbar}|\Psi(0)\rangle. \tag{3.3}$$

The dynamics of a mixed state is described by a time-dependent linear, positive, normalized and locally finite function $\rho(t, A)$ on the operator algebra, for which the local density matrices $\hat{\rho}_V(t)$, associated with the state $\rho(t, A)$ via (2.23), obey the equation of motion

$$-\frac{\hbar}{i}\frac{\partial}{\partial t}\hat{\rho}_V(t) = \hat{H}_\mu\hat{\rho}_V(t) - \hat{\rho}_V(t)\hat{H}_\mu \tag{3.4}$$

with the solution

$$\hat{\rho}_V(t) = e^{-i\hat{H}_\mu t/\hbar}\hat{\rho}_V(0)e^{i\hat{H}_\mu t/\hbar}. \tag{3.5}$$

This follows directly from the definition (2.17) of a density matrix and from (1.7). Observe the difference in sign compared to (1.8) and to (3.7) below, which is due to the fact that $\hat{\rho}_V$ is a state rather than an observable.

If an explicitly time-dependent field is applied to the considered system, the Hamiltonian may parametrically depend on time in these expressions even in the Schrödinger picture.[18]

The time-dependence of expectation values (2.15, 2.23) of local operators $\hat{A}$ localized in V is given by

$$\begin{aligned}\langle A\rangle(t) &= \rho(t,A) = \operatorname{tr}\left(\hat{\rho}_V(t)\hat{A}\right) = \operatorname{tr}\left(e^{-i\hat{H}_\mu t/\hbar}\hat{\rho}_V(0)e^{i\hat{H}_\mu t/\hbar}\hat{A}\right) = \\ &= \operatorname{tr}\left(\hat{\rho}_V(0)e^{i\hat{H}_\mu t/\hbar}\hat{A}e^{-i\hat{H}_\mu t/\hbar}\right) = \operatorname{tr}\left(\hat{\rho}_V\hat{A}(t)\right) = \rho(A(t)).\end{aligned} \quad (3.6)$$

Only the cyclicity of the trace of a product was used between the first and the second line, mediating the transition from the Schrödinger picture to the Heisenberg picture, where now the state ρ (as a function on the operator algebra) or $|\Psi\rangle$ (as a Hilbert space vector) is independent of time and instead the Heisenberg operators

$$\hat{A}(t) = e^{i\hat{H}_\mu t/\hbar}\hat{A}(0)e^{-i\hat{H}_\mu t/\hbar} \quad (3.7)$$

as solutions of (1.8) contain all time-dependences. In quantum field theory, calculations are preferably carried out on an algebraic level focusing on the operators rather than on states. States are usually built, out of a fixed given cyclic vector, by applying field operators. This causes a preference of the Heisenberg picture against the Schrödinger picture.

However, for interacting quantum systems, the time-dependence of Heisenberg operators may be very involved. Therefore, in most cases one splits the Hamiltonian according to

$$\hat{H}_\mu = \hat{H}_0 + \hat{H}_1 \quad (3.8)$$

into a solvable part $\hat{H}_0$ (which contains the term $-\mu\hat{N}$) and the rest, containing the interaction. Then, one introduces an interaction picture by *defining* a time-dependence of states according to

$$\hat{\rho}_{IV}(t) \stackrel{\text{def}}{=} e^{i\hat{H}_0 t/\hbar}\hat{\rho}_V(t)e^{-i\hat{H}_0 t/\hbar}. \quad (3.9)$$

---

[18]This is a pragmatic point of view for adiabatic situations where the time-dependence of the external field is sufficiently slow; the quantization under time-dependent side conditions is an unsolved problem of quantum physics.

The index I at $\hat{\rho}_{IV}(t)$ indicates the interaction picture, while $\hat{\rho}_V(t)$ as previously means a state in Schrödinger picture. We have

$$\langle A \rangle(t) = \operatorname{tr}\left(\hat{\rho}_V(t)\hat{A}\right) = \operatorname{tr}\left(e^{-i\hat{H}_0 t/\hbar}\hat{\rho}_{IV}(t)e^{i\hat{H}_0 t/\hbar}\hat{A}\right) =$$
$$= \operatorname{tr}\left(\hat{\rho}_{IV}(t)e^{i\hat{H}_0 t/\hbar}\hat{A}e^{-i\hat{H}_0 t/\hbar}\right) = \operatorname{tr}\left(\hat{\rho}_{IV}(t)\hat{A}_I(t)\right) \quad (3.10)$$

with the simple time-dependence of operators

$$\hat{A}_I(t) = e^{i\hat{H}_0 t/\hbar}\hat{A}e^{-i\hat{H}_0 t/\hbar} \quad (3.11)$$

in the interaction picture. Inserting (3.5) into the right-hand site of (3.9), one finds the time evolution operator

$$\hat{U}(t,0) = e^{i\hat{H}_0 t/\hbar}e^{-i\hat{H}_\mu t/\hbar} \quad (3.12)$$

of states in the interaction picture:

$$\hat{\rho}_{IV}(t) = \hat{U}(t,0)\hat{\rho}_{IV}(0)\hat{U}^\dagger(t,0), \quad \hat{\rho}_{IV}(0) = \hat{\rho}_V(0) = \hat{\rho}_V. \quad (3.13)$$

This time evolution operator $\hat{U}(t,0)$ is a complicated construct, since in general $\hat{H}_0$ and $\hat{H}_1$ do not commute.

3.2 Time-dependent Green's Functions

We are now going to study the time evolution of a quasi-stationary excitation at zero temperature. Let, at the beginning, the system be in its ground state.—Hier stock' ich schon![19]...—Recall that we want to create local excitations by means of field operators. The representation of field operators, however, includes the definition of the state space on which they act, preferably the Fock space with the cyclic state vector $|\Psi_0\rangle$, in our context meaning the state without local excitations. Of course, at least if we look at a system at zero temperature, $|\Psi_0\rangle$ may often mean the ground state. However, one and the same assembly of atoms may be found in different 'ground-states' at $T=0$ (for instance a solid consisting of carbon atoms may be found as diamond or as graphite; as was explained in Section 2.3, both cases correspond to canonical states in different irreducible representation spaces). Nevertheless, the true globally stable ground state with the absolute minimum of

---
[19] Goethe, Faust I, verse 1225.

the total energy can always be supposed unique with respect to its structure (it is not, e.g. with respect to the orientation of crystallographic axes in space). More complicated is the situation, if we consider a solid at $T > 0$. As was already mentioned in the introduction, for instance the phonons created in this case do not refer to the ground state but to an enlarged lattice spacing, depending on the numbers of phonons already present. This reference state without quasi-particles does not exist in reality. We shall come back to these questions in Chapter 6, and then shall finally define the state $|\Psi_0\rangle$. In order to have a definite notion, in the present chapter the reader may preliminarily assume $|\Psi_0\rangle$ to be the ground state of the system.

Hence, let, at the beginning, the system be in a state $|\Psi_0\rangle$ (in the Heisenberg picture) where quasi-stationary excitations are absent, and let it, at the time $t = 0$, make a transition into the state $|\Psi\rangle$ by creating such an excitation:

$$t = 0: \qquad |\Psi\rangle_{t=0} = \hat{\Phi}^\dagger_\alpha(t=0)|\Psi_0\rangle. \tag{3.14}$$

where $\hat{\Phi}^\dagger_\alpha(t)$ is a certain creation operator in Heisenberg picture, depending on parameters α (momentum or position, polarization, and maybe others).[20] This creation operator $\hat{\Phi}^\dagger_\alpha(t)$ is meant to be built up from field operators (1.140) or (1.149), its actual expression depends on the type of excitation considered. Examples will be given subsequently. As another idealization, we will preferably consider excitations with a given momentum, created by an operator of the type (1.149) where the r-dependence of $g(x)$ is given by the plane wave (1.52). This is of course not a local function in the sense of (1.149). It has to be understood as the main Fourier component of a really local wave pocket $g(x)$, which, however, is supposed to have such a large extension in space, that only one Fourier component need be considered with any wanted accuracy. As long as we are interested in the properties of the system within some finite volume V, we may take this volume to be the torus (1.53) and hence may consider the values (1.54) to be the admissible wave-vectors of momentum eigenstates.

The result of operating with a time-dependent Heisenberg operator on a time-independent state yields a state $|\Psi\rangle_t$ with a patametric time-dependence inherited from the Heisenberg operator and reversed to that of the Schrödinger picture. If

---

[20] See Chapter 5 for the connection of the parameters α with $|\Psi_0\rangle$. If $|\Psi_0\rangle$ has a low symmetry (e.g. in an amorphous solid), then this connection is not very essential: there is a rather large arbitrariness in the definition of elementary excitations. The quasi-particle picture is less effective in this case than for a crystal, but nevertheless it may turn out useful also here, e.g., to consider long-wavelength phonons.

3.2 Time-dependent Green's Functions

$|\Psi\rangle_0 = |\Psi\rangle_{t=0}$ were stationary, then, at $t > 0$, the system would up to a phase factor be found in the same state:

$$t > 0: \quad |\Psi\rangle_t = \hat{\Phi}_\alpha^\dagger(t)|\Psi_0\rangle \stackrel{\text{stationary}}{=} |\Psi\rangle_0 e^{i\xi_\alpha t/\hbar} \qquad (3.15)$$

with a *real* excitation energy $\xi_\alpha > 0$. Since $|\Psi\rangle$ is only quasi-stationary, it decays in time (with some decay time τ_α),

$$|\Psi\rangle_t \approx |\Psi\rangle_0 e^{i\xi_\alpha t/\hbar - t/\tau_\alpha} + |\Psi'\rangle_t, \quad {}_t\langle\Psi'|\Psi\rangle_0 = 0, \qquad (3.16)$$

and the probability

$$w_\alpha(t) = |{}_t\langle\Psi|\Psi\rangle_0|^2 = |\langle\Psi_0|\hat{\Phi}_\alpha(t)\hat{\Phi}_\alpha^\dagger(0)|\Psi_0\rangle|^2 \qquad (3.17)$$

for the system to be found in the state (3.14) at $t > 0$ is smaller than unity.

If $|\Psi_0\rangle$ was an N-particle state of the system, and $\hat{\Phi}_\alpha^\dagger$ creates one particle, then $|\Psi\rangle$ is an $(N+1)$-particle state. As is well known, an excited state with $N-1$ particles can also be formed out of $|\Psi_0\rangle$: by acting on it with an annihilation operator $\hat{\Phi}_\alpha$ and thus forming a 'hole' in an orbital which was occupied in $|\Psi_0\rangle$. In order to treat both cases simultaneously by one single expression, the latter action is supposed to have taken place at some time $t < 0$ and its result is observed at $t = 0$, hence

$$\begin{aligned}
w_\alpha(|t|) &= |G_\alpha(t)|^2, \\
G_\alpha(t) &= -i\langle\Psi_0|\mathbf{T}\hat{\Phi}_\alpha(t)\hat{\Phi}_\alpha^\dagger(0)|\Psi_0\rangle \stackrel{\text{def}}{=} \\
&\stackrel{\text{def}}{=} \begin{cases} -i\langle\Psi_0|\hat{\Phi}_\alpha(t)\hat{\Phi}_\alpha^\dagger(0)|\Psi_0\rangle & t > 0 \\ \mp i\langle\Psi_0|\hat{\Phi}_\alpha^\dagger(0)\hat{\Phi}_\alpha(t)|\Psi_0\rangle & \text{for} \quad t < 0, \end{cases}
\end{aligned} \qquad (3.18)$$

where in the last line of the expression for the *Green's function* $G_\alpha(t)$ the sign '−' stands in the case of a boson operator $\hat{\Phi}_\alpha$ and the sign '+' stands in the case of a fermion operator. By definition, the time-ordering operator $\mathbf{T}$ puts the Heisenberg operators right to it into an order of descending times from left to right, and applies statistics sign factors in case of permutation of operators. (The rest of the pre-factor is convention.)

This *ordinary* Green's function observes the rules of causality in consistency with the 'particle content' of the state $|\Psi_0\rangle$: it considers that a particle, not present in $|\Psi_0\rangle$, must be created *before* it can be annihilated, and that a hole must be created *before* it can be filled again. Therefore, it is sometimes called the *causal* Green's function. Besides, for *technical and conceptual* reasons, 'retarded' and 'advanced' Green's functions are introduced and used in many-body techniques (see next section). They are particularly useful at non-zero temperatures where they have simpler analytical properties (the reason why they are used), and their names are connected with the technical definition of their time-dependence rather then with physical aspects. Only if $|\Psi_0\rangle$ is the *vacuum*, the retarded Green's function coincides with the ordinary (causal) one. Since it describes the propagation of excitations, the Green's function is also called a propagator.

(Exercise. Show $G_\alpha(t) = -i\langle\Psi_{I0}(t)|\hat{\Phi}_{I\alpha}(t)\hat{U}(t,0)\hat{\Phi}^\dagger_{I\alpha}(0)|\Psi_{I0}(0)\rangle$ for $t > 0$ and a corresponding relation for $t < 0$. The subscript I means interaction picture.)

3.3 Non-Interacting Particles

In a system of *non-interacting* particles we would have $\xi_\alpha = \varepsilon_\alpha - \mu$ in (3.15), where ε_α is the energy of the created particle corresponding to the Hamiltonian $\hat{H}$. The again positive excitation energy of a hole excitation would be $\xi_\alpha = \mu - \varepsilon_\alpha$ (see Fig.4 on page 86), while the pre-factor would change the sign according to the lower sign of the lower line of (3.18). Hence, in this case, for a *fermion* operator $\hat{\Phi}_\alpha$,

$$G_\alpha(t) = \begin{cases} -ie^{-i(\varepsilon_\alpha-\mu)t/\hbar} & \text{for} \quad t > 0,\ \varepsilon_\alpha > \mu \\ +ie^{-i(\varepsilon_\alpha-\mu)t/\hbar} & \quad\quad\quad t < 0,\ \varepsilon_\alpha < \mu \\ 0 & \text{elsewhere.} \end{cases} \quad (3.19)$$

A Fourier transformation

$$G_\alpha(\omega) = \int_{-\infty}^{\infty} dt\, e^{i\omega t/\hbar} G_\alpha(t) \quad (3.20)$$

of the expression (3.19) yields

$$G_\alpha(\omega \pm i0) = \frac{\hbar}{\omega - \varepsilon_\alpha + \mu \pm i0} \quad \text{for} \quad \varepsilon_\alpha \gtrless \mu. \quad (3.21)$$

An infinitesimal imaginary part has to be added to the frequency ω in the indicated manner in order to obtain a converging Fourier integral in (3.20). By convention, we measure frequencies ω in energy units, so that $\omega/\hbar$ is an ordinary frequency in units of inverse time.

For a *boson* operator $\hat{\Phi}_\alpha$, the second line in (3.19) is to be omitted since there are no bosons to be annihilated in the state Ψ_0 of a non-interacting bosonic system, with the only possible exception that these bosons are in the Bose condensate.

3.3 Non-Interacting Particles

Annihilation of a boson from the condensate, however, would not result in an excited state $\Psi(t)$ in (3.14) (it would merely yield the $(N-1)$-particle ground state), so that we everywhere in the following presume α to be such that $\hat{\Phi}_\alpha$ does not correspond to a condensate particle. Having this in mind we obtain for a system of non-interacting bosons[21]

$$G_\alpha(\omega + i0) = \frac{\hbar}{\omega - \varepsilon_\alpha + \mu + i0}, \qquad \varepsilon_\alpha > \mu \tag{3.22}$$

for the frequency-dependent (Fourier transformed) Green's function.

For a *finite* system in a *finite volume*, the real excitation energies ε_α are discretely distributed and nowhere clustering. Hence, the Green's function of a finite system is a meromorphic function of the complex frequency variable ω (a function having discretely distributed poles and being otherwise analytic). This result holds true for interacting finite systems in finite volumes, too. This is easily seen by casting the Green's function into the Lehmann representation: insert a completeness relation of stationary many-particle states between the two field operators of (3.18) and pass over to the Schrödinger picture, make the time-dependence of the stationary Schrödinger states explicit and Fourier transform each item (exercise).[22]

The poles in the complex ω-plane of this meromorphic function are placed above the real axis for Re $\omega < 0$ and below the real axis for Re $\omega > 0$. In the case of an interacting system there is a great many of such poles (with small residua),

---

[21] In condensed matter theory, the majority of bosons may separately be created (not necessarily in pairs). For them always $\mu = 0$ holds [Landau and Lifshits, 1980a, §63]. Besides, it is convenient to introduce the Hermitian field operator

$$\hat{A}_\alpha = \sqrt{\hbar/2\varepsilon_\alpha}(\hat{\Phi}_\alpha + \hat{\Phi}_\alpha^\dagger)$$

for them, and to introduce the corresponding Green's function

$$D_\alpha(t) = -i\langle\Psi_0|\mathbf{T}\hat{A}_\alpha(t)\hat{A}_\alpha(0)|\Psi_0\rangle.$$

For it, the well known representation

$$D_\alpha(\omega) = \frac{\hbar}{2\varepsilon_\alpha}\left[\frac{\hbar}{\omega - \varepsilon_\alpha + i0} - \frac{\hbar}{\omega + \varepsilon_\alpha - i0}\right] = \frac{\hbar^2}{\omega^2 - (\varepsilon_\alpha - i0)^2}$$

is obtained (exercise). (Note that in $D_\alpha(\omega)$ both poles correspond to *the same positive* excitation energy.) For our present purposes, however, the expressions (3.18) and (3.22) are quite sufficient.

[22] This Lehmann representation or spectral representation of the Green's function of an interacting system is only of formal theoretical use because of its enormous number of terms; see also Section 2.4.

and in the thermodynamic limit these poles merge into cut lines, infinitesimally set off from the negative and positive parts of the real ω-axis, respectively. Since the real ω-axis intersects the cut line, there is no analytic continuation of the Green's function from the whole real axis.

In order to prepare for exploiting the theory of complex functions, two new Green's functions

$$G_\alpha^{r/a}(t) \stackrel{\text{def}}{=} \mp i\theta(\pm t)\langle \Psi_0 | \left[\hat{\Phi}_\alpha(t), \hat{\Phi}_\alpha^\dagger(0) \right]_\zeta | \Psi_0 \rangle \qquad (3.23)$$

are introduced. θ is the step function. The superscripts 'r' and 'a' mean retarded and advanced, respectively, the upper signs are belonging to the retarded function and the lower to the advanced function. The (anti-)commutator is to be used in accordance with statistics, that is, $\zeta = \pm$ for fermions and bosons, respectively. Now, it is easily seen, either for the non-interacting case or in general using the Lehmann representation, that the retarded Green's function is analytic in the whole upper half plane and the advanced Green's function is analytic in the whole lower half plane of the complex ω-plane. Even more easily this follows from

$$G_\alpha^r(t) = \begin{cases} G_\alpha(t) - G_\alpha(-t) & \text{for} \quad t > 0 \\ 0 & t < 0 \end{cases} \qquad (3.24)$$

and

$$G_\alpha^a(t) = \begin{cases} 0 & \text{for} \quad t > 0 \\ G_\alpha(t) - G_\alpha(-t) & t < 0 \end{cases}, \qquad (3.25)$$

which is readily deduced from the definitions of all three Green's functions (exercise).

A complex function being analytic in one half plane is called a Herglotz function. Its limes on the real axis may even be a distribution, the simplest example of such a situation being given by Dyson's formula

$$\frac{1}{\omega - \omega_0 \pm i0} = \wp \frac{1}{\omega - \omega_0} \mp i\pi\delta(\omega - \omega_0) \qquad (3.26)$$

for real ω and ω_0. The symbol $\wp$ means the principal value, and the left and right expressions of the equation are understood as factors of the integrand of a contour integral in the ω-plane. (Distributions, like the δ-function, are always understood as factors under an integral sign.)

3.4 Interacting Systems

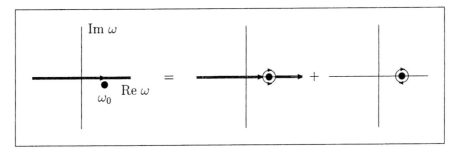

Figure 3: Contour integration along the real axis of the complex ω-plane. Dyson's formula (3.26, case of upper signs) holds for the integrand of this integral and hence for any deformation of the integration path in the complex plane.

For instance, for a non-interacting fermion system we have now

$$G_\alpha^r(\omega + i0) = \frac{\hbar}{\omega - \varepsilon_\alpha + \mu + i0} \quad \text{for all } \varepsilon_\alpha, \tag{3.27}$$

and combining this with (3.26), we may write for the density of states $D(\varepsilon)$ of single-particle energies ε

$$D(\varepsilon) = -\frac{1}{\pi\hbar} \operatorname{Im} \sum_\alpha G_\alpha^r(\varepsilon) = \sum_\alpha \delta(\varepsilon - \varepsilon_\alpha). \tag{3.28}$$

In a matrix generalization of Green's functions according to

$$G_{\alpha\beta}(t) = -i\langle \Psi_0 | \mathbf{T}\hat{\Phi}_\alpha(t)\hat{\Phi}_\beta^\dagger(0) | \Psi_0 \rangle \tag{3.29}$$

this density of states may be written as

$$D(\varepsilon) = -\frac{1}{\pi\hbar} \operatorname{Im} \operatorname{tr} G^r(\varepsilon). \tag{3.30}$$

This formula and its modifications will subsequently be used to *define* the density of states of quasi-stationary excitations by way of analogy.

3.4 Interacting Systems

The formulas (3.19–3.30) refer to a quantum gas of noninteracting particles. They are given here, because the theory of quasi-stationary excitations in macrosystems is build using the theory of quantum gases as a prototype. If now, at $t = 0$, $\hat{\Phi}_\alpha^\dagger(0)$ does not create a stationary state $|\Psi\rangle_0$ but only a quasi-stationary one, then, at $t > 0$, $|\Psi\rangle_t$ is no longer given by the last expression of (3.15). Instead, this contribution to $|\Psi\rangle_t$ will decay in time, probably according to the law (3.16) with the lifetime τ_α of the quasi-stationary state. Consequently the Green's function (3.18) also contains (among others) terms (3.19) with certain coefficients, but with *complex* energies ε_α, where, because the decay proceeds with increasing $|t|$,

$$\operatorname{Im} \varepsilon_\alpha = -\frac{\hbar}{\tau_\alpha}\operatorname{sign}(\operatorname{Re} \varepsilon_\alpha - \mu). \tag{3.31}$$

This means that $G_\alpha(\omega)$ contains a term[23]

$$G_\alpha(\omega) = \frac{\hbar Z_\alpha}{\omega - \varepsilon_\alpha + \mu} + \cdots, \tag{3.32}$$

and therefore

$$G_\alpha^{-1}(\varepsilon_\alpha - \mu) = 0. \tag{3.33}$$

This equation, together with the definitions (3.18) and (3.20) gives the complex energy ε_α of a quasi-stationary excitation as a function of the parameters α [Abrikosov et al., 1975, Landau and Lifshits, 1980b].[24]

---

[23] Our mere phenomenological considerations do not indicate the occurrence of a factor Z_α in (3.32) different from unity. The point is that in actual situations the creation operator of the quasi-particle state $\hat{\Phi}_\alpha^{q\dagger}|\Psi_0\rangle$ is not known. There are usually only indications that a given field operator $\hat{\Phi}_\alpha$ contains the quasi-particle operator $\hat{\Phi}_\alpha^q$ as a constituting part,

$$\hat{\Phi}_\alpha = \sqrt{Z_\alpha}\hat{\Phi}_\alpha^q + \text{something}.$$

For instance in a metal the quasi-particle 'conduction electron' consists of a 'bare electron' and a surrounding it 'polarization cloud'. (In the axiomatic field theory one tries to derive the expression for $\hat{\Phi}_\alpha^q$ and to calculate the renormalization constant Z_α of the field amplitude. In applied theories the renormalization procedures have always a phenomenological character, since they are based on choosing special classes of diagrams and so on.) In an actual theory the Green's function (3.18) may only be build up with given field operators $\hat{\Phi}_\alpha$ expressed through the operators entering the Hamiltonian. These expressions may in special cases be complicated by the demand that the $\hat{\Phi}_\alpha$ by *definition* be fermionic or bosonic [Akhieser et al., 1967].

[24] As was mentioned, for a finite system, $G_\alpha(\omega)$ is a meromorphic function and hence defined on the whole complex ω-plane except for isolated points. In the thermodynamic limit these points merge into cut lines infinitesimally above the negative real ω-axis and below the positive

3.4 Interacting Systems

Strictly speaking, no proof can be given for the decay of the quasi-stationary excitations to follow a mere exponential law (3.16), but it can be expected that in an appropriately chosen interval of time, $\Delta t \sim \tau_\alpha$, the expression (3.16) describes the decay in a sufficiently precise manner. As a consequence, the Green's function $G_\alpha(\omega)$ has a more complicated structure compared to (3.32), on the real axis of the complex ω-plane it may, however, be well approximated by the expression (3.32). Non-exponential relaxation processes in interacting systems, not entering the dispersion relations and lifetimes of quasi-particles are physically real. They become essential in cases not adequately described by quasi-particles. It should be noted in this connection that the whole quasi-particle conception, based on the analytical structure of (3.32), and as we shall see below, also any particle concept in quantum field theory rests on this *realistic assumption* suggested by experimental facts with the help of theoretical reasoning along the indicated lines (compare the very clear treatment in [Anderson, 1997, Chapter 3]).

The restriction in the choice of the field operators $\hat{\Phi}_\alpha$ in general to be of the Bose or Fermi type is again dictated by phenomenology. In the relativistic field theory, this restriction results in the well-known manner from the causality principle in space-time, also called the 'locality principle', under the additional presumption that the (anti-)commutator of the field operators be a c-number; again a requisite of phenomenology [Bogolubov and Shirkov, 1959, Haag, 1993]. In the case of condensed matter, the state $|\Psi_0\rangle$ has a lower symmetry compared to Poincaré invariance (invariance with respect to inhomogeneous Lorentz transformations, the symmetry of the vacuum in relativistic quantum field theory), and in principle operators different from a pure Bose or Fermi type can be useful. For instance in the theory of magnetism spin operators $\hat{S}^\dagger$ and $\hat{S}$ are used, which only in the special case $S = 1/2$ (in which case we denoted them $\hat{\sigma}^\dagger$ and $\hat{\sigma}$ in (1.158) at a given site l) have fermionic properties. For large S they attain approximate bosonic properties. If, however, these operators do not 'reduce' to a Bose or Fermi type at least in the limiting case of small occupation numbers, they cannot be utilized to set up an independent-particle picture as the starting approximation for treating the interaction by means of perturbation theory.

---

real ω-axis. The original ω-plane, cut in this way, is called the physical sheet, and by analytic continuation of $G_\alpha(\omega)$ through the cut this physical sheet can be continued into a Riemann surface. The pole of (3.32), for instance for $\varepsilon_\alpha > \mu$, is obtained by continuation of $G_\alpha(\omega)$ from the positive real axis to *negative* values of Im ω as seen from (3.31), and hence is on the 'unphysical' part of that Riemann surface.

Of course, for a quasi-stationary excitation,

$$|\operatorname{Im} \varepsilon_\alpha| \ll |\operatorname{Re} \varepsilon_\alpha - \mu| \tag{3.34}$$

is supposed. In solids, in most actual situations, this condition is fulfilled for not too high[25] excitation energies $|\operatorname{Re} \varepsilon_\alpha - \mu|$ and, moreover, these excitations are exhaustive in the sense that at sufficiently low temperature the grand canonical partition function of the macrosystem may be expressed as

$$\begin{aligned}Z_{\text{tot}}(T,V) &= \left(\sum_{\{n_\alpha\}} \exp\left[-\sum_\alpha n_\alpha \xi_\alpha(V)/T\right]\right) Z_{\text{int}}(T,V) = \\ &= \left(\prod_\alpha [1 \mp e^{-\xi_\alpha(V)/T}]^{\mp 1}\right) Z_{\text{int}}(T,V),\end{aligned} \tag{3.35}$$

where the upper and lower signs correspond to bosons and fermions, respectively, and

$$\xi_\alpha \approx |\operatorname{Re} \varepsilon_\alpha - \mu| \tag{3.36}$$

is (approximately; see Chapter 4) the excitation energy. Here, one must have in mind that not only the energies ξ_α themselves depend on the volume V, but also the parameters α. (Usually this latter dependence is rather trivial: for instance, if the summation over the quasi-momenta is replaced by an integration, a factor V in the quasi-particle density of states is produced via (1.55).) The correction factor $Z_{\text{int}}(T,V)$ describes the interaction of the excitations (e.g. anharmonicity). In the simplest cases it tends to unity, if $T \to 0$, or the smallness of $\ln Z_{\text{int}}$ corresponds to a hierarchy of interactions (see, e.g., [Akhieser et al., 1967]).

In situations described by (3.34-3.36) the knowledge of the spectrum of stationary states with energies E_i is already not necessary to calculate the various thermodynamic quantities at sufficiently low temperatures. It suffices to know the *spectra of elementary quasi-stationary excitations* in the just indicated sense of (3.35) with energies ξ_α and, maybe, the knowledge of their mutual interactions[26].

---

[25]There are indications that in certain cases too low excitation energies must also be excluded in view of (3.34), for instance in the case of electronic excitations in disordered solids [Shklovskii and Efros, 1984].

[26]It must again be stressed that it is in this connection *necessary*, that the quasi-stationary excitations 'exhaust' all possible motions with energies $\sim T$ in the system. The knowledge of the

As was mentioned in connection with the definition of a Green's function by (3.18), in the case of a fermion operator $\hat{\Phi}_\alpha$, besides the quasi-stationary excited states $\hat{\Phi}_\alpha^\dagger|\Psi_0\rangle$ of particle type with excitation energies Re $\varepsilon_\alpha - \mu > 0$ and ε_α being determined by the equation (3.33), there are also states $\hat{\Phi}_\alpha|\Psi_0\rangle$ of the hole type with excitation energies $\mu - $ Re $\varepsilon_\alpha > 0$ and ε_α again being obtained from (3.33). Just for this reason appear the signs of the absolute value in (3.36). According to their construction the particle and hole type excitations have opposite inner characteristics (charges) and may therefore mutually annihilate emitting one or several neutral bosons. Since, however, generally speaking, in a solid there is no analogue of the external CPT symmetry of elementary particle theory [Bogolubov and Shirkov, 1959, Haag, 1993], the external characteristics as excitation energies (or masses) and scattering or reaction cross-sections of particles and holes are completely independent from each other.

A simple example of the described situation is a spatially homogeneous Fermi liquid as ^{3}He in the normal state. Due to the homogeneity, here the excitation has a definite momentum. The fermion particle-hole excitation spectrum is shown on Fig.4a, next page. In view of (3.36), it may be expressed as

$$\text{Re } \varepsilon(p) - \mu = v_f\,(p - p_f) = \frac{p_f}{m^*}\,(p - p_f), \tag{3.37}$$

$$\text{Im } \varepsilon(p) \sim (\text{Re } \varepsilon(p) - \mu)^2, \tag{3.38}$$

where p is the absolute value of the momentum, and p_f is the Fermi momentum (see below in Chapter 5). Formally (3.37) is analogous to the spectrum of an interaction-free fermion gas (Fig.4b, see also page 27f), however, the 'effective mass' m^* exceeds the mass of a free ^{3}He atom approximately by a factor of 3. (This increase of mass is caused by the participation of all He atoms in the motion with momentum p.) The various types of sound waves form the boson branches of the elementary excitation spectra of liquid ^{3}He [Landau and Lifshits, 1980b, Khalatnikov, 1985].

3.5 Density of States

If a quasi-particle spectrum $\varepsilon_\alpha = \xi_\alpha + i\hbar/\tau_\alpha$ has been determined via (3.33) (or in any equivalent way or just by guess or modeling), then, for instance to treat the

---

sound-wave spectra in a classical liquid does not allow to determine its free energy, because the sound waves do not exhaust all motions in the liquid: the atoms also move quasi-independently dragging in only the nearest neighbors. Just this motion hardly submits to a simple description, thus making a quantitative theory of a classical liquid so difficult.

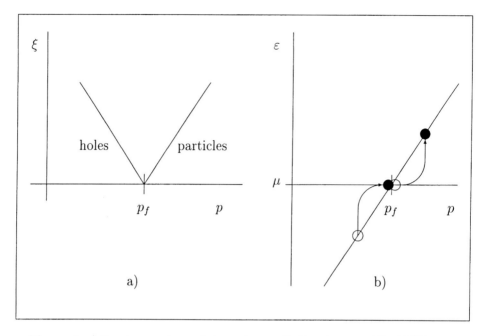

Figure 4: a) Fermion spectrum of elementary excitations in normal liquid ^{3}He. b) Energy spectrum of an interaction-free fermion gas. All particle states with energies $\varepsilon < \mu$ are occupied in the ground state. A one-particle excitation is obtained either by lifting a particle from below μ up to the Fermi level μ leaving a hole, or by lifting the particle from the level μ to a higher one, the excitation energy in both cases being positive and equal to $|\varepsilon - \mu|$, and the excited state deviates from an $(N \pm 1)$-particle ground state only by the occupation of a *single* orbital.

thermodynamics via (3.35), it is convenient to introduce a *quasi-particle density of states* as

$$D_{\rm qp}(\varepsilon) = \sum_\alpha \delta(\varepsilon - \varepsilon_\alpha) \qquad (3.39)$$

like in (3.28). Also, if stationary transport is considered, for instance via the quantum Boltzmann equation (using the Fermi distribution and Pauli's exclusion principle in the scattering integral), it has to be taken into account that one quasi-electron (quasi-hole) carries exactly one quantum of negative (positive) charge. The polarization cloud around a quasi-particle does not transport charge in a stationary situation. (It does of course contribute to high-frequency currents.) Again, the quasi-particle density of states or its momentum resolved analogue is a key quantity.

This is to be contrasted, for instance to tunneling processes, which are proportional to the square of the excitations $\hat{\Phi}^\dagger|\Psi_0\rangle$ on both sides of the barrier or junction. These excitations are produced by injection of a primary particle, not a quasi-particle since quasi-particles do not tunnel as an unperturbed entity. Compared to the quasi-particle, the 'primary-particle content' of these states is reduced by a factor $\sqrt{Z_\alpha}$, the quasi-particle spectral amplitude renormalization constant. Hence, these processes are governed by another density of states which we prefer to call *spectral density of states* but which is often (somewhat misleadingly) called single-particle density of states in the literature. Comparing (3.32) with (3.30) reveals that it is the spectral density of states which is given by (3.30) and that one can put

$$D_{\text{qp}}(\varepsilon) = -\frac{1}{\pi\hbar} \operatorname{Im} \operatorname{tr}\left(Z^{-1}(\varepsilon) G^{\text{r}}(\varepsilon)\right). \tag{3.40}$$

The relevance of these two different densities of states again gives evidence of the approximative character of the notion of quasi-particle. For non-interacting particles both densities of states are equal ($Z = 1$).

Another case where causion of this type is needed is the long-range assymptotics of the potential produced by a charged quasi-particle, a quasi-electron say, in a solid, that is, the assymptotics of quasi-electron quasi-electron interaction. Since the range of the polarization cloud is the same as that of the potential, it contributes to the assymptotics of the latter. It turns out [Kohn, 1958, Ambegaokar, 1961] that this assymptotics is

$$v(\boldsymbol{r}) \to \frac{e}{\epsilon r}, \tag{3.41}$$

where e is the electrostatic charge of the electron (entire charge quantum) and ϵ is the static dielectric constant of the solid. Hence, in a semiconductor or insulator ($1 < \epsilon < \infty$) the long-range interaction is that of reduced effective charges, and in a metal ($\epsilon = \infty$) the quasi-electron interaction falls off more rapidly than $1/r$.

3.6 Collective Excitations

The quasi-stationary excitations considered so far were excitations with raising the total particle number in the system by one (particle excitations) or with lowering it by one (hole excitations). A simultaneous particle-hole excitation, which does not change the total particle number in the system, is not elementary in the sense that when both the excited particle and hole move into different remote parts of the system, then the excitation energy and other quantum numbers like momentum, spin and so on are *sums* of the individual quantum numbers of the particle and of the hole. However, another important type of *elementary* excitations in condensed matter goes without change of the total particle number. These are certain 'eigenmodes of vibration' of the system (and hence bosonic in nature).

Suppose that there is a time-dependent external field $F(x,t)$, which couples to some density $\hat{n}_\alpha(x)$ of our system according to a perturbative part of the Hamiltonian in the Schrödinger picture

$$\hat{H}'(t) = \int dx\, \hat{n}_\alpha(x) F(x,t). \tag{3.42}$$

Note that the time-dependence here is parametric through the external field. As an example, think of an electron field $\hat{\psi}(x)$ and an external electric potential $U(x,t)$ varying so weakly in time that the corresponding magnetic field can be neglected. Then, $\hat{H}'(t) = \int dx\, \hat{\psi}^\dagger(x)(-e)\hat{\psi}(x) U(x,t)$.

Let $\hat{H}'(t<0) = 0$ and $|\Psi(t=0)\rangle = |\Psi_0\rangle$. Put, in the Schrödinger picture and for $t > 0$, the perturbed state $|\Psi'(t)\rangle$ in the form

$$|\Psi'(t)\rangle = e^{-i\hat{H}_\mu t/\hbar} \hat{V}(t) |\Psi_0\rangle, \quad \hat{V}(0) = 1. \tag{3.43}$$

$\hat{H}_\mu$ is the Hamiltonian without external perturbation (but with interaction in the system). From the perturbed Schrödinger equation

$$-\frac{\hbar}{i}\frac{\partial}{\partial t}|\Psi'(t)\rangle = \left[\hat{H}_\mu + \hat{H}'(t)\right]|\Psi'(t)\rangle \tag{3.44}$$

one finds for the perturbation of the time evolution

$$-\frac{\hbar}{i}\frac{\partial \hat{V}}{\partial t} = e^{i\hat{H}_\mu t/\hbar}\hat{H}'(t) e^{-i\hat{H}_\mu t/\hbar} \hat{V}(t) \tag{3.45}$$

with the iterative solution, obtained by first putting $\hat{V}(t) \approx 1$ on the right side and integrating, (then putting the result for $\hat{V}(t)$ on the right side again and so on,)

$$\hat{V}(t) = 1 - \frac{i}{\hbar}\int_0^t dt'\, e^{i\hat{H}_\mu t'/\hbar} \hat{H}'(t') e^{-i\hat{H}_\mu t'/\hbar} + \cdots \tag{3.46}$$

3.6 Collective Excitations

The lowest order density perturbation is now (exercise, use (3.43) and (3.46))

$$\begin{aligned}\langle \delta n_\alpha(x,t)\rangle &= \langle \Psi'(t)|\hat{n}_\alpha(x)|\Psi'(t)\rangle - \langle \Psi_0(t)|\hat{n}_\alpha(x)|\Psi_0(t)\rangle = \\ &= \langle \Psi_0|\frac{i}{\hbar}\int_0^t dt'\left[e^{i\hat{H}_\mu t'/\hbar}\hat{H}'(t')e^{-i\hat{H}_\mu t'/\hbar},\ \hat{n}_\alpha(x,t)\right]_-|\Psi_0\rangle = \\ &= \frac{i}{\hbar}\int_0^t dt'\int dx'\,\langle \Psi_0|\,[\hat{n}_\alpha(x',t'),\ \hat{n}_\alpha(x,t)]_-\,|\Psi_0\rangle F(x',t'),\end{aligned}$$
(3.47)

where we turned back to the Heisenberg picture in the second line. This second line is the well known Kubo formula for the linear response.

The (linear) susceptibility $\chi_{\alpha\alpha}(x,t;x',t')$ relates the linear density response to the perturbing potential,

$$\langle \delta n_\alpha(x,t)\rangle = \int_{-\infty}^\infty dt'\int dx'\chi_{\alpha\alpha}(x,t;x',t')F(x',t'). \tag{3.48}$$

Comparing this relation with our last result we see that the susceptibility

$$\chi_{\alpha\alpha}(x,t;x',t') = -\frac{i}{\hbar}\theta(t-t')\langle \Psi_0|\,[\hat{n}_\alpha(x,t),\ \hat{n}_\alpha(x',t')]_-\,|\Psi_0\rangle \tag{3.49}$$

is a retarded Green's function. (The time integral was extended to $-\infty$ without harm because $F(x',t')$ in the integrand was zero for negative t'; in this notation the result remains valid if the perturbation is switched on at any time.) Because the unperturbed Hamiltonian $\hat{H}_\mu$ was time-independent, the susceptibility depends on the time difference $t - t'$ only.

Fourier transformation of the time-dependence yields

$$\langle \delta n_\alpha(x,\omega)\rangle = \int dx'\,\chi_{\alpha\alpha}(x,x';\omega)F(x',\omega). \tag{3.50}$$

If now $\chi_{\alpha\alpha}$ as a function of ω has a pole, then *a density perturbation can propagate in the system without an external field*. This is the anticipated collective mode. Examples are the plasmon in a metal, the zero sound in liquid ^{3}He, phonons and magnons in crystalline solids (the latter at least for low occupation numbers).

Since χ is a bosonic Green's function, these excitations are bosons. Moreover, since χ is a retarded Green's function, it is Herglotz, that is, analytic in the upper half plane of the complex ω-plane. Thus, the poles can only be below the real ω-axis, and the collective modes have a finite lifetime.

3.7 Non-Zero Temperatures

The time-dependent Green's functions at non-zero temperature are defined in complete analogy to the case $T = 0$, only the state $|\Psi_0\rangle$ is to be replaced by the *canonical* state (2.31), or particularly with a complete set of stationary many-particle states, by (2.29):

$$G_\alpha(t) = -i \operatorname{tr}\left(\hat{\rho}_{T,V} \mathbf{T}\hat{\Phi}_\alpha(t)\hat{\Phi}_\alpha^\dagger(0)\right), \tag{3.51}$$

$$G_\alpha^{\mathrm{r/a}}(t) = \mp i\theta(\pm t) \operatorname{tr}\left(\hat{\rho}_{T,V}\left[\hat{\Phi}_\alpha(t), \hat{\Phi}_\alpha^\dagger(0)\right]_\zeta\right). \tag{3.52}$$

Their Lehmann representation (for *finite* volume) is obtained in the usual manner by inserting completeness relations of stationary states between the factors under the trace of (3.51) and passing over to the Schrödinger picture:

$$
\begin{aligned}
G_\alpha(t) &= -i \sum_{mnp} \langle p| \frac{1}{Z(V,T)} e^{-\hat{H}_\mu/T} |m\rangle \cdot \\
&\quad \cdot \left(\theta(t)\langle m|\hat{\Phi}_\alpha(t)|n\rangle\langle n|\hat{\Phi}_\alpha^\dagger(0)|p\rangle \pm \theta(-t)\langle m|\hat{\Phi}_\alpha^\dagger(0)|n\rangle\langle n|\hat{\Phi}_\alpha(t)|p\rangle\right) = \\
&= -i \sum_{mn} \frac{1}{Z(V,T)} e^{-(E_m(V)-\mu N_m)/T} \cdot \\
&\quad \cdot \left(\theta(t)e^{-i\omega_{nm}t/\hbar}\langle m|\hat{\Phi}_\alpha|n\rangle\langle n|\hat{\Phi}_\alpha^\dagger|m\rangle \pm \theta(-t)e^{i\omega_{nm}t/\hbar}\langle m|\hat{\Phi}_\alpha^\dagger|n\rangle\langle n|\hat{\Phi}_\alpha|m\rangle\right).
\end{aligned}
\tag{3.53}
$$

The upper sign in the parentheses is for bosons and the lower one for fermions, and we introduced a shorthand notation

$$\omega_{nm} = E_n(V) - E_m(V) - \mu(N_n - N_m). \tag{3.54}$$

By interchanging the summation indices of the second contribution, the result may be cast into

$$
\begin{aligned}
G_\alpha(t) &= -i \sum_{mn} \frac{1}{Z(V,T)} e^{-(E_m(V)-\mu N_m)/T} |\langle m|\hat{\Phi}_\alpha|n\rangle|^2 e^{-i\omega_{nm}t/\hbar} \cdot \\
&\quad \cdot \left(\theta(t) \pm \theta(-t)e^{-\omega_{nm}/T}\right),
\end{aligned}
\tag{3.55}
$$

3.7 Non-Zero Temperatures

with the Fourier transform

$$G_\alpha(\omega) = \hbar \sum_{mn} \frac{1}{Z(V,T)} e^{-(E_m(V)-\mu N_m)/T} |\langle m|\hat{\Phi}_\alpha|n\rangle|^2 \cdot$$
$$\cdot \left[\wp \frac{1}{\omega - \omega_{nm}}(1 \pm e^{-\omega_{nm}/T}) - i\pi \delta(\omega - \omega_{nm})(1 \mp e^{-\omega_{nm}/T}) \right] \qquad (3.56)$$

as the final Lehmann representation.

The difference in the temperature factors at the real and imaginary parts in the brackets of (3.56) reflects the fact that these expressions result by means of Dyson's formula (3.26) from pole terms which are densely distributed on both sides of the whole real ω-axis. Hence, in the thermodynamic limit this Green's function can again nowhere be analytically continued into the complex ω-plane. Nevertheless, from (3.56) the relation

$$\text{Re } G_\alpha(\omega) = \frac{\wp}{\pi} \int_{-\infty}^\infty d\epsilon \frac{\text{Im } G_\alpha(\epsilon)}{\epsilon - \omega} \tanh^\zeta \frac{\epsilon}{2T} \qquad (3.57)$$

is easily obtained.

By contrast, using (3.24) and (3.25) one finds directly from (3.56) and (3.57) (now, the two signs refer to r and a while ζ is further on the statistics factor)

$$G_\alpha^{r/a}(\omega) = \text{Re } G_\alpha(\omega) \pm i\text{Im } G_\alpha(\omega) \tanh^\zeta \frac{\omega}{2T} \qquad (3.58)$$

and the *dispersion relations*

$$\text{Re } G_\alpha^{r/a}(\omega) = \pm \frac{\wp}{\pi} \int_{-\infty}^\infty d\epsilon \frac{\text{Im } G_\alpha^{r/a}(\epsilon)}{\epsilon - \omega}, \qquad (3.59)$$

$$\text{Im } G_\alpha^{r/a}(\omega) = \mp \frac{\wp}{\pi} \int_{-\infty}^\infty d\epsilon \frac{\text{Re } G_\alpha^{r/a}(\epsilon)}{\epsilon - \omega}, \qquad (3.60)$$

which imply that $G_\alpha^{r/a}(\omega)$ are analytic in the upper/lower half-planes of the complex ω-plane. This Herglotz character can of course immediately be deduced from the definition (3.52). In the thermodynamic limit the poles of the Lehmann representation coalesce again into cut lines, and, for instance *the retarded Green's function, can again be analytically continued into the lower half plane (of an 'unphysical' sheet), and its poles there again yield the quasi-stationary excitations.*

Nevertheless, due to the appearance of the full Hamiltonian in both the temperature and time-evolution terms of (3.51) and (3.52), none of those Green's functions allows a comprehensible perturbation expansion which could be formalized into a diagrammatic technique.

We are not really concerned with diagrammatic techniques in this volume. There exist excellent treatises of this subject, for instance the classic [Abrikosov et al., 1975] or [Landau and Lifshits, 1980b]. We sketch only in brief the connection with the retarded Green's function, which latter contains the information on the quasi-particle spectra. The entity suited for a diagrammatic treatment is the Matsubara Green's function defined as

$$\mathcal{G}_\alpha(\tau) = -iG_\alpha(-i\hbar\tau) = -\operatorname{tr}\left(\hat{\rho}_{T,V}\mathbf{T}_\tau\hat{\Phi}_\alpha(-i\hbar\tau)\hat{\Phi}_\alpha^\dagger(0)\right). \tag{3.61}$$

It is taken at 'imaginary time' t, $\tau = it/\hbar$, $-\beta \leq \tau \leq \beta = 1/T$, and uses the analytic continuation (2.32) of a Heisenberg operator. The 'imaginary-time ordering operator' $\mathbf{T}_\tau$ orders the factors in a sequence with descending real values τ in the negative-imaginary arguments, with statistics sign changes in commutations of factors understood. From the KMS relation (2.33) it follows immediately that

$$\mathcal{G}_\alpha(-\tau) = \pm\mathcal{G}_\alpha(\beta - \tau), \qquad \tau > 0, \tag{3.62}$$

and hence its Fourier series

$$\mathcal{G}_\alpha(\tau) = \frac{1}{\beta}\sum_n e^{-i\omega_n\tau}\mathcal{G}_\alpha(\omega_n) \tag{3.63}$$

has non-zero Fourier coefficients

$$\mathcal{G}_\alpha(\omega_n) = \frac{1}{2}\int_{-\beta}^{\beta} d\tau\, e^{i\omega_n\tau}\mathcal{G}_\alpha(\tau) \tag{3.64}$$

only for

$$\omega_n = \begin{cases} 2n\pi/\beta & \text{Bosons} \\ (2n+1)\pi/\beta & \text{Fermions} \end{cases} \tag{3.65}$$

and with

$$\mathcal{G}_\alpha(-\omega_n) = \mathcal{G}_\alpha^*(\omega_n). \tag{3.66}$$

3.7 Non-Zero Temperatures

The Lehmann representation is easily calculated to be

$$\mathcal{G}_\alpha(\omega_n) = \hbar \sum_{mn} \frac{1}{Z(V,T)} e^{-(E_m(V) - \mu N_m)/T} |\langle m|\hat{\Phi}_\alpha|n\rangle|^2 \cdot$$
$$\cdot \frac{1}{i\omega_n - \omega_{nm}}(1 \pm e^{-\omega_{nm}/T}). \tag{3.67}$$

Hence,

$$G_\alpha^r(i\omega_n) = \mathcal{G}_\alpha(\omega_n), \qquad \omega_n \geq 0 \tag{3.68}$$

(only G_α^r is defined on the upper half-plane), and under the constraint $G_\alpha^r(\omega) \to 0$ for $|\omega| \to \infty$, $G_\alpha^r(\omega)$ can be reconstructed by analytical continuation from the non-negative Matsubara frequencies onto its whole domain of analyticity. (Likewise, G_α^a can be reconstructed from its values at non-positive Matsubara frequencies.)

We finally just explain why the Matsubara Green's function undergoes a diagrammatic technique. Consider the time evolution operator (3.12) of the interaction picture at $t = -i\hbar\beta$,

$$\hat{U}(-i\hbar\beta, 0) = e^{\beta \hat{H}_0} e^{-\beta \hat{H}_\mu}, \quad \text{i.e.} \quad e^{-\beta \hat{H}_\mu} = e^{-\beta \hat{H}_0} \hat{U}(-i\hbar\beta, 0). \tag{3.69}$$

The trace of the last expression is just $Z(T,V)$. With the help of these relations the Matsubara Green's function (3.61) for $0 \leq \tau \leq \beta$ may be rewritten in the interaction picture as

$$\mathcal{G}_\alpha(\tau) =$$
$$= -\frac{\text{tr}\left(e^{-\beta \hat{H}_0} \hat{U}(-i\hbar\beta, 0)\hat{U}(0, -i\hbar\tau)\hat{\Phi}_{I\alpha}(-i\hbar\tau)\hat{U}(-i\hbar\tau, 0)\hat{\Phi}_{I\alpha}^\dagger(0)\right)}{\text{tr}\left(e^{-\beta \hat{H}_0}\hat{U}(-i\hbar\beta, 0)\right)}$$
$$= -\frac{\text{tr}\left(e^{-\beta \hat{H}_0} \hat{U}(-i\hbar\beta, -i\hbar\tau)\hat{\Phi}_{I\alpha}(-i\hbar\tau)\hat{U}(-i\hbar\tau, 0)\hat{\Phi}_{I\alpha}^\dagger(0)\right)}{\text{tr}\left(e^{-\beta \hat{H}_0}\hat{U}(-i\hbar\beta, 0)\right)}$$
$$= -\frac{\langle \mathbf{T}_\tau \hat{\Phi}_{I\alpha}(-i\hbar\tau)\hat{\Phi}_{I\alpha}^\dagger(0)\hat{U}(-i\hbar\beta, 0)\rangle_0}{\langle \hat{U}(-i\hbar\beta, 0)\rangle_0}. \tag{3.70}$$

The group property of $\hat{U}$ was used, and the unperturbed expectation value is defined as

$$\langle \hat{A} \rangle_0 = \frac{\text{tr}\left(e^{-\beta \hat{H}_0}\hat{A}\right)}{\text{tr}\left(e^{-\beta \hat{H}_0}\right)}. \tag{3.71}$$

Under the $\mathbf{T}_\tau$ operator, the factors in the numerator of the last expression for $\mathcal{G}_\alpha$ were rearranged, and the $\hat{U}$-factors were collected together at the end. For $\tau < 0$ the same result is analogously obtained. This last expression of (3.70) is precisely the general basis of a diagrammatic expansion.

Thermodynamic quantities (macro-variables) may directly be obtained from the Matsubara Green's function without the detour via analytically continuing it to the retarded Green's function.

4 Model Hamiltonians

Condensed matter theory deals with model Hamiltonians rather than with the 'Hamiltonian of the real world' whatever that actually could be. Generally, the models to be used vary with density (or alternatively chemical potential), with temperature, with volume (at least with dimensionality), and so on.

The object of this chapter is to show how a Hamiltonian depending on the thermodynamic variables comes into the game, to which extent it can be modeled, and how it can approximately be derived in simple situations. The central issue is that the primary Hamiltonian as the operator of the total energy can be split into a part commuting with the whole operator algebra and which hence, in view of (1.8), does not affect the local dynamics, and a part that determines this dynamics and that survives the thermodynamic limit. In particular, the first part provides the free energy (thermodynamic potential) of the thermodynamic equilibrium state, and consequently the second part depends on the parameters of that state. On the level of mathematical physics, this splitting problem has been finally solved in the famous paper by [Haag et al., 1967]. Here, we will not follow these lines but rather illustrate the situation on the basis of explicit comprehensible examples (see also [Thirring, 1980]): non-interacting particles at zero and non-zero temperature T, and models of Fermi and Bose systems with interactions simple enough to be tractable with canonical transformations of the field operators.

In this chapter we essentially deal with finite systems, for instance confined in the volume V or periodically moving in a torus (1.53), and only occasionally consider the thermodynamic limit.

4.1 Non-Interacting Particles, $T = 0$

The grand canonical Hamiltonian of an interaction-free fermion gas may be written as

$$\hat{H}_\mu = \sum_\alpha \hat{c}_\alpha^\dagger (\varepsilon_\alpha - \mu) \hat{c}_\alpha, \tag{4.1}$$

where the ε_α are the single-particle energies in single-particle energy eigenstates ϕ_α created by $\hat{c}_\alpha^\dagger$, and the $\hat{c}$-operators obey the canonical equal-time anticommutation relations (1.94). The ground state $|\Psi_0\rangle$ has all orbitals ϕ_α with $\varepsilon_\alpha < \mu$ occupied and all orbitals with $\varepsilon_\alpha > \mu$ unoccupied, and hence is characterized by

$$\hat{c}_\alpha^\dagger |\Psi_0\rangle = 0 \text{ for } \varepsilon_\alpha < \mu, \quad \hat{c}_\alpha |\Psi_0\rangle = 0 \text{ for } \varepsilon_\alpha > \mu. \tag{4.2}$$

If the particles carry a conserved charge q, then

$$\hat{Q} = \sum_\alpha \hat{c}_\alpha^\dagger q \hat{c}_\alpha \tag{4.3}$$

is the operator of the total charge in the system, which commutes with the Hamiltonian as demanded by charge conservation.

Instead of the $\hat{c}$-operators, new fermion operators $\hat{\zeta}$, $\hat{\xi}$ may be introduced according to

$$\hat{c}_\alpha = \begin{cases} \hat{\zeta}_\alpha \\ s_\alpha \hat{\xi}_{\bar{\alpha}}^\dagger \end{cases} \text{ for } \varepsilon_\alpha \begin{matrix} > \\ < \end{matrix} \mu \tag{4.4}$$

and a Hermitian conjugate relation for $\hat{c}_\alpha^\dagger$. Here, s_α is a constant phase factor, $|s_\alpha|^2 = 1$, and $\bar{\alpha}$ is related to α by inversion of the particle orbital motion and spin quantum numbers.[27] Specifically, if α classifies with respect to the projection of the total angular momentum of the particle, m_α, in a given direction and S is the (half integer) total angular momentum itself, then

$$s_\alpha = (-1)^{S-m_\alpha}, \quad m_{\bar{\alpha}} = -m_\alpha, \quad \boldsymbol{p}_{\bar{\alpha}} = -\boldsymbol{p}_\alpha. \tag{4.5}$$

We skip here the detailed justification of the choice for s_α (see, for instance, [Fetter and Walecka, 1971, §56]).

Each of the two sets of new fermion operators $\hat{\zeta}$, $\hat{\xi}$ obviously obeys canonical anticommutation relations analogous to (1.94) for the $\hat{c}$-operators while all operators of one set anticommute with all operators of the other. Therefore the operator transformation (4.4) is called a canonical transformation. For $\varepsilon_\alpha > \mu$, $\hat{\zeta}_\alpha^\dagger$ creates a particle like $\hat{c}_\alpha^\dagger$, but for $\varepsilon_\alpha < \mu$, $\hat{\xi}_{\bar{\alpha}}^\dagger$ creates a hole.

The former ground state $|\Psi_0\rangle$ is the Fock space vacuum $|\rangle$ of the new operators:

$$\hat{\zeta}_\alpha |\rangle = 0, \quad \hat{\xi}_\alpha |\rangle = 0 \tag{4.6}$$

---

[27]This means that the orbitals ϕ_α and $\phi_{\bar{\alpha}}$ are related to each other by 'time reversal'. Note that in the absence of magnetic fields or spontaneous spin polarization time reversal symmetry is realized and hence $\varepsilon_{\bar{\alpha}} = \varepsilon_\alpha$ even if the solid has no spatial inversion symmetry. Therefore, in this case $\varepsilon_\alpha < \mu$ implies $\varepsilon_{\bar{\alpha}} < \mu$.

4.1 Non-Interacting Particles, $T = 0$

for all α for which the $\hat{\zeta}$, $\hat{\xi}$-operators are defined. Note that $|\Psi_0\rangle$ and $|\rangle$ are only different notations for the same state. There is no danger of confusion because the vacuum of the $\hat{c}$-operators will not at all be used in this chapter.

Replacement of the $\hat{c}$-operators by the $\hat{\zeta}$, $\hat{\xi}$-operators in the Hamiltonian and in the operator of total charge, and transformation to normal order in the new operators results in

$$\hat{H}_\mu = \sum_\alpha^{\varepsilon_\alpha < \mu}(\varepsilon_\alpha - \mu) + \sum_\alpha^{\varepsilon_\alpha < \mu} \hat{\xi}^\dagger_{\bar{\alpha}}|\varepsilon_\alpha - \mu|\hat{\xi}_{\bar{\alpha}} + \sum_\alpha^{\varepsilon_\alpha > \mu} \hat{\zeta}^\dagger_\alpha(\varepsilon_\alpha - \mu)\hat{\zeta}_\alpha =$$
$$= \Omega(\mu) + \hat{H}^p_\mu, \qquad (4.7)$$

$$\hat{Q} = \sum_\alpha^{\varepsilon_\alpha < \mu} q + \sum_\alpha^{\varepsilon_\alpha < \mu} \hat{\xi}^\dagger_{\bar{\alpha}}(-q)\hat{\xi}_{\bar{\alpha}} + \sum_\alpha^{\varepsilon_{alpha} > \mu} \hat{\zeta}^\dagger_\alpha q \hat{\zeta}_\alpha =$$
$$= Q(\mu) + \hat{Q}^p_\mu. \qquad (4.8)$$

The Hamiltonian now consists of a c-number term, which is equal to the grand canonical thermodynamic potential $\Omega = -pV$ as a function of the chemical potential μ (at temperature $T = 0$; p is the pressure), and of a *positive definite* operator term $\hat{H}^p_\mu$, which contains the hole and particle excitation energies rather then the former particle energies (cf. Fig. 4). The operator of total charge consists of a c-number term, equal to the charge in thermodynamic equilibrium determined by μ, and an operator part, accounting that particles and holes have opposite charges.

The original quantum field (1.140) is now expressed as

$$\hat{\psi}(x) = \sum_\alpha^{\varepsilon_\alpha < \mu} s_\alpha \phi_\alpha(x) \hat{\xi}^\dagger_{\bar{\alpha}} + \sum_\alpha^{\varepsilon_\alpha > \mu} \phi_\alpha(x)\hat{\zeta}_\alpha \qquad (4.9)$$

and consists of a hole part and a particle part. In the hole part, the creator $\hat{\xi}^\dagger_{\bar{\alpha}}$ is attached with the original orbital ϕ_α. However, if for instance ϕ_α was a plane wave with momentum $\boldsymbol{p}_\alpha$, then $\phi_\alpha(x) \sim \exp(i\boldsymbol{p}_\alpha \cdot \boldsymbol{r}/\hbar) = \exp(-i\boldsymbol{p}_{\bar{\alpha}} \cdot \boldsymbol{r}/\hbar) = [\exp(i\boldsymbol{p}_{\bar{\alpha}} \cdot \boldsymbol{r}/\hbar)]^* \sim \phi^*_{\bar{\alpha}}$, that is, the hole created by $\hat{\xi}^\dagger_{\bar{\alpha}}$ carries a momentum $\boldsymbol{p}_{\bar{\alpha}} = -\boldsymbol{p}_\alpha$ which is the correct momentum of the excitation if a particle carrying the momentum $\boldsymbol{p}_\alpha$ is removed from the system. If ϕ_α was a Bloch wave of local parts having a local orbital momentum, then, expressed in terms of the local azimuthal variable φ, $\phi_\alpha \sim e^{im_{l\alpha}\varphi}$, and from (4.5) again $\phi_\alpha \sim \phi^*_{\bar{\alpha}}$. Finally, if ϕ_α as a spin orbital is represented by a spinor like (1.17) for the spin 1/2 case, then there exists a spin matrix operator $\hat{C}$ so that $s_\alpha \phi_\alpha = \hat{C}\phi_{\bar{\alpha}}^*$. From (4.5) and (1.10) it is easily seen that for the spin 1/2 case $C = -i\hat{\sigma}_y$. Therefore, (4.9) may be rewritten as

$$\hat{\psi}(x) = \sum_\alpha^{\varepsilon_\alpha < \mu} \hat{C}\phi_{\bar{\alpha}}^*(x)\hat{\xi}_{\bar{\alpha}}^\dagger + \sum_\alpha^{\varepsilon_\alpha > \mu} \phi_\alpha(x)\hat{\zeta}_\alpha, \qquad (4.10)$$

which has the same form as the electron-positron field in Quantum Electrodynamics (e.g. [Itzykson and Zuber, 1980, Eqs. (3.157), (3.185) and (2.97a)]). Our choice in (4.4) ensures that a hole created by $\hat{\xi}_{\bar{\alpha}}^\dagger$ has motional quantum numbers α as the quantum numbers of the excitation of the $(N-1)$ particle state left behind the removed particle.

Let us elaborate a little bit more on that latter point. Consider an isotropic medium (for instance liquid ^{3}He) with an energy momentum relation $\varepsilon = p^2/(2m^*)$ of the particles, determined by an effective mass $m^* > 0$. The group velocity of a wave pocket formed from momentum eigenstates with eigenvalues close to $\boldsymbol{p}_\alpha$ is $\boldsymbol{v}_\alpha = (\partial \varepsilon/\partial \boldsymbol{p})_\alpha = \boldsymbol{p}_\alpha/m^*$. The dispersion relation for the excitation energy of hole excitations is now $\varepsilon = (p_f^2 - p^2)/(2m^*)$ which leads to $\boldsymbol{v}_\alpha = -\boldsymbol{p}_\alpha/m^* = \boldsymbol{p}_\alpha/(-m^*)$. In the considered case the hole excitation has a negative effective mass: if the wave pocket of the hole (wave pocket of the removed particle) travels to the right, the increment of total momentum of the excited system after the particle removal is to the left. (See text before (4.10).) The sign of mass of the hole is opposite to that of the removed particle.[28] All inner quantum numbers of a hole excitation as mass, charge, magnetic moment as well as their motional quantum numbers as momentum and angular momentum are opposite to those of the removed particle; hence the gyromagnetic ratio which is the ratio between the magnetic moment and the angular momentum is the same for a hole and the corresponding removed particle, with *equal sign*. This ratio is usually expressed as $g(q/2m^*)$ where q is the charge and g is the relative gyromagnetic ratio. Since the ratio q/m^* is also the same for a hole and the corresponding removed particle, g is also equal for both.

Now, recall the twofold rôle of the Hamiltonian in quantum physics: as the operator of total energy and as the generator of time evolution according to (1.8). In the latter respect, $\hat{H}_\mu$ and $\hat{H}_\mu^p$ are completely equivalent, since a c-number commutes with every operator and hence does not influence the Heisenberg equations of motion. (In the Schrödinger picture, an additive c-number Ω of the Hamiltonian is completely compensated by a physically ineffective gauge transformation

---

[28]In Quantum Electrodynamics the 'electron in the Dirac sea' has negative mass and the positron has positive mass. In condensed matter physics negative masses of either electrons or hole excitations are rather the rule than the exception; see next chapter.

4.1 Non-Interacting Particles, $T = 0$

$|\Psi(t)\rangle \to |\Psi(t)\rangle e^{-i\Omega t/\hbar}$.) However, while $\hat{H}_\mu$ diverges in the thermodynamic limit and only

$$\lim_{V \to \infty} \frac{\hat{H}_\mu}{V} = \lim_{V \to \infty} \frac{\Omega(\mu)}{V} = -p(\mu) \tag{4.11}$$

retains a meaning, $\hat{H}_\mu^p$ remains well defined and continues to describe the (local) dynamics, since its expectation value in the vacuum is zero with zero fluctuations. The same holds true for $\hat{Q}_\mu^p$, while $\hat{Q}/V$ yields the thermodynamic charge density.

The Hamiltonian of an interaction-free boson gas, expressed through the operators of single-particle energy eigenstates, is analogously

$$\hat{H}_\mu = \sum_\alpha \hat{b}_\alpha^\dagger (\varepsilon_\alpha - \mu) \hat{b}_\alpha, \tag{4.12}$$

where for a finite particle number N the chemical potential must be below the lowest single-particle energy ε_0 and it approaches this value for $T \to 0$. Since the chemical potential fixes the macroscopic expectation value of the particle number rather than its eigenvalue, and since for $T = 0$, i.e. $\mu = \varepsilon_0$ the ground states for all N (with only the lowest orbital occupied by N particles) are degenerate with respect to $\hat{H}_\mu$, we choose a (normalized to unity) coherent state rather than an N-particle state as the ground state. Put $b_0 = \sqrt{N}$, $b_{\alpha \neq 0} = 0$ in (1.102) and take

$$|\Psi_0\rangle = \exp\left(\sqrt{N} \hat{b}_0^\dagger\right) |0\rangle e^{-N/2}, \tag{4.13}$$

where we renamed the vacuum of (1.102) by $|0\rangle$ reserving the notation $|\rangle$ for later purpose. In view of (1.105), $|\Psi_0\rangle$ is normalized to unity. Considering (1.98) and (1.107) yields

$$\begin{aligned} \hat{b}_0 |\Psi_0\rangle &= |\Psi_0\rangle \sqrt{N}, \\ \hat{b}_0^\dagger |\Psi_0\rangle &= \left(\frac{\partial}{\partial \sqrt{N}} |\Psi_0\rangle e^{N/2}\right) e^{-N/2} \approx |\Psi_0\rangle \sqrt{N}. \end{aligned} \tag{4.14}$$

The state $|\Psi_0\rangle$ is a linear combination of states $|n_0 000 \ldots\rangle$ (in the notation (1.84)) with N-dependent coefficients smaller than unity in absolute value, and peaking around $n_0 = N$. Consider for instance (1.101) and (1.102). Now, a little reflection will make it clear that the derivative of $|\boldsymbol{b}\rangle = |\Psi_0\rangle \exp(N/2)$ with respect to $\sqrt{N}$ is dominated by that of the exponential factor, and for $N \to \infty$ the last relation (4.14) becomes rigorous.

Thus we may replace both $\hat{b}_0$ and $\hat{b}_0^\dagger$ by the same c-number $\sqrt{N}$ and hence remove them from the set of generators of the operator algebra, and may accordingly take $|\rangle = |\Psi_0\rangle$ as the vacuum of the remaining algebra:

$$\hat{b}_\alpha |\rangle = 0 \text{ for all } \alpha \neq 0. \tag{4.15}$$

Furthermore, we have

$$\hat{H}_\mu = \sum_{\alpha \neq 0} \hat{b}_\alpha^\dagger (\varepsilon_\alpha - \mu) \hat{b}_\alpha = \Omega(\mu) + \hat{H}_\mu^p, \tag{4.16}$$

where $\Omega(\mu) = 0$. It has been written down for reasons of analogy only. There is no pressure at $T = 0$: the system is Bose condensed and the condensate has no pressure. If the particles carry a charge q, we have for the operator of total charge

$$\hat{Q} = qN + \sum_{\alpha \neq 0} \hat{b}_\alpha^\dagger q \hat{b}_\alpha. \tag{4.17}$$

The original quantum field is expressed as

$$\hat{\psi}(x) = \phi_0(x)\sqrt{N} + \sum_{\alpha \neq 0} \phi_\alpha(x) \hat{b}_\alpha. \tag{4.18}$$

In contrast to the fermion case, the separation of the Bose condensate $|\phi_0(x)|^2 N$ from $\hat{\psi}^\dagger(x)\hat{\psi}(x)$ is not a canonical transformation, and it is only rigorous in the thermodynamic limit.

4.2 Non-Interacting Particles, $T > 0$

The approach of the last section may be generalized to the case of non-zero temperature $T > 0$. (In the bosonic case we drop the special treatment for temperatures below the Bose condensation temperature T_B and restrict the consideration to $T > T_B$, that is, $\mu < \varepsilon_0$.) To prepare for this generalization, we first represent the thermodynamic state as a vector state in a reducible representation of the operator algebra instead of the usual density matrix representation (2.17) with $p_i = (e^{(\varepsilon_\alpha - \mu)/T} \pm 1)^{-1}$ for fermions/bosons.

For a finite system, a reducible vector representation of any mixed state can be realized as a direct product of the Fock space with itself. If $\{|\Psi_i\rangle\}$ is an orthonormal

4.2 Non-Interacting Particles, $T > 0$

basis in the Fock space $\mathcal{F}$ which diagonalizes the density matrix as in (2.17), then a vector of the new space $\mathcal{F} \otimes \mathcal{F}$ is given by

$$|\Psi\rangle\rangle \stackrel{\text{def}}{=} |\Psi^{(1)} \otimes \Psi^{(2)}\rangle = \sum_{ik} |\Psi_i \otimes \Psi_k\rangle C_{ik}. \tag{4.19}$$

The Fock space operators $\hat{A}$ are represented by

$$\hat{\hat{A}} \stackrel{\text{def}}{=} \hat{A} \otimes 1 = \sum_{iklm} |\Psi_i \otimes \Psi_k\rangle A_{im} \delta_{kl} \langle \Psi_l \otimes \Psi_m|, \tag{4.20}$$

where the bra-vector $\langle \Psi_l \otimes \Psi_m|$ is defined as $|\Psi_m \otimes \Psi_l\rangle^\dagger$. The operators act as an identity in the second factor of the representation space leaving it invariant (whence the representation is reducible). Hence, we have

$$\langle\langle \Psi | \hat{\hat{A}} | \Psi \rangle\rangle = \sum_k \sum_{im} C_{ik}^* A_{im} C_{mk}. \tag{4.21}$$

Now, define the mixed state vector to be

$$|P\rangle\rangle = \sum_i |\Psi_i \otimes \Psi_i\rangle \sqrt{p_i}, \quad \text{i.e.} \quad C_{ik} = \sqrt{p_i} \delta_{ik}, \tag{4.22}$$

then

$$\langle\langle P | \hat{\hat{A}} | P \rangle\rangle = \sum_k p_k A_{kk} = \operatorname{tr}_F \hat{\rho} \hat{A}. \tag{4.23}$$

As we see, in our reducible representation space, the normalized vector $|P\rangle\rangle$ describes the same state over operators $\hat{\hat{A}}$ as the density matrix $\hat{\rho}$ describes over operators $\hat{A}$ in the Fock space (cf. (2.17)).

A particular choice of a basis in the Fock space was the occupation number basis (1.84), in which the ground state of non-interacting particles, $|\Psi_0\rangle$ of last section, was a single basis state. By a unitary basis transformation of both vectors and operators in $\mathcal{F} \otimes \mathcal{F}$, it can likewise be achieved that the thermodynamic state at temperature T, volume V and chemical potential μ is a single basis state. After such a unitary transformation, the operators $\hat{\hat{A}}$ have already no longer the form (4.20). Of course, the reducibility of the representation remains unaffected by this unitary equivalence.

For the non-interacting particle system, within the latter frame we introduce the operators

$$\hat{\hat{a}}_\alpha = (\mp 1)^{\hat{N}} \otimes \frac{\hat{a}^\dagger_\alpha}{\sqrt{e^{(\varepsilon_\alpha-\mu)/T} \pm 1}} + \frac{\hat{a}_\alpha}{\sqrt{1 \pm e^{-(\varepsilon_\alpha-\mu)/T}}} \otimes 1 \qquad (4.24)$$

Again, the upper sign is for fermions and the lower sign for bosons. The operators $\hat{a}_\alpha$ (and $\hat{a}^\dagger_\alpha$) in the numerators of the right hand side are ordinary Fock space operators, $\hat{c}_\alpha$ for fermions and $\hat{b}_\alpha$ for bosons, with the properties reported in Subsection 1.2.1. With those properties it is easily verified (exercise) that

$$[\hat{\hat{a}}_\alpha, \hat{\hat{a}}^\dagger_\beta]_\pm = \delta_{\alpha\beta}, \qquad [\hat{\hat{a}}_\alpha, \hat{\hat{a}}_\beta]_\pm = 0 = [\hat{\hat{a}}^\dagger_\alpha, \hat{\hat{a}}^\dagger_\beta]_\pm . \qquad (4.25)$$

The factor $(-1)^{\hat{N}}$ with the ordinary Fock space operator $\hat{N}$ of the total particle number in the first factor of $\mathcal{F} \otimes \mathcal{F}$ is essential to ensure in the fermionic case that the cross products of the both terms of (4.24) cancel in the anticommutators (4.25). We see that the operators (4.24) together with their Hermitian conjugates provide a (reducible) representation of the canonical (anti-)commutation relations.

In quantum mechanics, probability amplitudes figure as wavefunction expansion coefficients, the squares of the absolute values of which are probabilities. The construction of $\hat{\hat{a}}^\dagger$ (Hermitian conjugate of (4.24)),

$$\hat{\hat{a}}^\dagger_\alpha = (\mp 1)^{\hat{N}} \otimes \frac{\hat{a}_\alpha}{\sqrt{e^{(\varepsilon_\alpha-\mu)/T} \pm 1}} + \frac{\hat{a}^\dagger_\alpha}{\sqrt{1 \pm e^{-(\varepsilon_\alpha-\mu)/T}}} \otimes 1, \qquad (4.26)$$

must take account of the probability of creating particles or holes. In the thermodynamic equilibrium state at $T > 0$, the probability of creating a hole is naturally proportional to the average number $\langle n_\alpha \rangle = (e^{(\varepsilon_\alpha-\mu)/T} \pm 1)^{-1}$ of *particles* present in that state. The probability of creating a particle on the other hand is in the fermionic case proportional to the average number of *holes* present in that state (a state must be empty in order to create a particle in it), and it consists in the case of bosons of a spontaneous and an induced part (cf. Einstein's theory of photon emission). This probability is proportional to $1 \mp \langle n_\alpha \rangle$. The square roots of both probability factors for hole and particle creation appear as factors at both terms of (4.26), respectively.

As already said, in the bosonic case, $\mu < \varepsilon_0$ is considered only. In the fermionic case, the $T \to 0$ limit of (4.24) is obviously

$$\hat{\hat{c}}_\alpha = \begin{cases} \hat{c}_\alpha \otimes 1 \\ (-1)^{\hat{N}} \otimes \hat{c}^\dagger_\alpha \end{cases} \hat{=} \begin{cases} \hat{\zeta}_\alpha \\ s_\alpha \hat{\xi}^\dagger_{\bar{\alpha}} \end{cases} \text{ for } \varepsilon_\alpha \begin{matrix} > \\ < \end{matrix} \mu. \qquad (4.27)$$

4.2 Non-Interacting Particles, $T > 0$

This compares to (4.4), but here we still have a representation in $\mathcal{F} \otimes \mathcal{F}$ while in (4.4) it was in $\mathcal{F}$.

As in an irreducible representation of the canonical (anti-)commutation relations, the relations (4.25) completely determine the algebraic structure of the algebra *generated by the operators* $\hat{\tilde{a}}_\alpha$, $\hat{\tilde{a}}^\dagger_\alpha$. For instance, the occupation number operator defined as previously has the properties

$$\hat{\tilde{n}}_\beta \stackrel{\text{def}}{=} \hat{\tilde{a}}^\dagger_\beta \hat{\tilde{a}}_\beta, \quad [\hat{\tilde{n}}_\beta, \hat{\tilde{a}}_\alpha]_- = -\delta_{\beta\alpha}\hat{\tilde{a}}_\alpha, \quad [\hat{\tilde{n}}_\beta, \hat{\tilde{a}}^\dagger_\alpha]_- = \delta_{\beta\alpha}\hat{\tilde{a}}^\dagger_\alpha. \tag{4.28}$$

However, while in an irreducible representation space like $\mathcal{F}$ the algebra generated by the annihilators and creators exhausts *all* linear operators of that space, in our reducible representation space there are plenty of operators not generated by the annihilators and creators of the physical system. In the representation (4.20) (which is not the same as (4.24); these two representations are only unitarily equivalent), all operators not of the form $\hat{A} \otimes 1$ are not generated by the annihilators and creators. Among those, all operators of the form $1 \otimes \hat{A}$ *commute* with the whole algebra (4.20). For instance, defining the operators

$$\hat{\tilde{\nu}}_\beta \stackrel{\text{def}}{=} -1 \otimes \hat{a}^\dagger_\beta \hat{a}_\beta + \hat{a}^\dagger_\beta \hat{a}_\beta \otimes 1,$$

the relations

$$[\hat{\tilde{\nu}}_\beta, \hat{\tilde{a}}_\alpha]_- = -\delta_{\beta\alpha}\hat{\tilde{a}}_\alpha, \quad [\hat{\tilde{\nu}}_\beta, \hat{\tilde{a}}^\dagger_\alpha]_- = \delta_{\beta\alpha}\hat{\tilde{a}}^\dagger_\alpha. \tag{4.29}$$

are easily demonstrated with the help of (4.24) (exercise). The operators $\hat{\tilde{\nu}}_\beta$ are not generated by (4.24). If we further define

$$\hat{\tilde{\mu}}_\beta \stackrel{\text{def}}{=} \hat{\tilde{n}}_\beta - \hat{\tilde{\nu}}_\beta, \quad \hat{\tilde{n}}_\beta = \hat{\tilde{\mu}}_\beta + \hat{\tilde{\nu}}_\beta, \tag{4.30}$$

we see immediately from (4.28, 4.29) that $\hat{\tilde{\mu}}_\beta$ commutes with the whole physical operator algebra generated by (4.24).

The Hamiltonian of the non-interacting particle system may now be cast into

$$\begin{aligned}\hat{\tilde{H}}_\mu &= \sum_\alpha (\varepsilon_\alpha - \mu)\hat{\tilde{n}}_\alpha = \\ &= \sum_\alpha (\varepsilon_\alpha - \mu)\hat{\tilde{\mu}}_\alpha - \\ &\quad - \sum_\alpha 1 \otimes \hat{a}^\dagger_\alpha(\varepsilon_\alpha - \mu)\hat{a}_\alpha + \sum_\alpha \hat{a}^\dagger_\alpha(\varepsilon_\alpha - \mu)\hat{a}_\alpha \otimes 1 = \\ &= \hat{\tilde{\Omega}}(T,V,\mu) + \hat{\tilde{H}}^p_\mu(T,V).\end{aligned} \tag{4.31}$$

Like the $\hat{\bar{\mu}}_\alpha$, the operator $\hat{\bar{\Omega}}(T,V,\mu)$ commutes with the whole algebra (4.24) and hence has no influence on the local quantum dynamics which is completely determined by $\hat{H}^p_\mu(T,V)$. In view of the structure of that operator we define a vacuum $|\rangle\rangle$ by

$$1 \otimes \hat{a}_\alpha |\rangle\rangle = 0, \quad \hat{a}_\alpha \otimes 1 |\rangle\rangle = 0 \quad \text{for all } \alpha. \tag{4.32}$$

One easily finds (exercise)

$$\langle\langle|\hat{\bar{n}}_\alpha|\rangle\rangle = \frac{1}{e^{(\varepsilon_\alpha-\mu)/T} \pm 1}, \quad \hat{\bar{\nu}}_\alpha|\rangle\rangle = 0, \tag{4.33}$$

and hence

$$\Omega(T,V,\mu) = \langle\langle|\hat{\bar{\Omega}}(T,V,\mu)|\rangle\rangle = \sum_\alpha \frac{\varepsilon_\alpha - \mu}{e^{(\varepsilon_\alpha-\mu)/T} \pm 1}, \tag{4.34}$$

which is the correct expression for the grand canonical thermodynamic potential of the interaction-free system. In the considered representation, the vacuum (4.32) is obviously the unitary equivalent to the state (4.22) with the grand canonical values p_i. Although this thermodynamic state $|\rangle\rangle$ is now an eigenstate of $\hat{H}^p_\mu(T,V)$ with zero eigenvalue and zero fluctuations, the energy of the system, expressed by the original Hamiltonian, of course thermodynamically fluctuates in the thermodynamic state. Also, $\hat{H}^p_\mu$ is not positive definite any more. This corresponds to the fact that in the thermodynamic state there is a non-zero probability to remove a particle from a state with energy arbitrarily high above the chemical potential.

A rigorous but formal splitting of the Hamiltonian into the energy of a density matrix state and a generator of time evolution which survives the thermodynamic limit for *any* system, interaction-free or not, has been given in [Haag *et al.*, 1967].

4.3 BCS Theory

In this section we treat an interacting fermion system by means of a canonical transformation. We consider a system of fermions with an isotropic energy-momentum dispersion relation ε^0_k, $k = |\boldsymbol{k}|$, with an attractive interaction between particles with opposite momentum and opposite spin so that there appears a possibility to form zero-momentum spin-singlet Cooper pairs. The Hamiltonian under consideration is

$$\hat{H}_\mu = \sum_{\boldsymbol{k}\sigma} \hat{c}^\dagger_{\boldsymbol{k}\sigma}(\varepsilon^0_k - \mu)\hat{c}_{\boldsymbol{k}\sigma} - \sum_{\boldsymbol{k}\boldsymbol{k}'} \hat{c}^\dagger_{\boldsymbol{k}'\uparrow}\hat{c}^\dagger_{-\boldsymbol{k}'\downarrow}V_{|\boldsymbol{k}'-\boldsymbol{k}|}\hat{c}_{-\boldsymbol{k}\downarrow}\hat{c}_{\boldsymbol{k}\uparrow} \tag{4.35}$$

4.3 BCS Theory

with $V_k > 0$.

In order to separate the ground state and the excitation spectrum, we perform the so-called Bogolubov-Valatin transformation [Bogolubov, 1958, Valatin, 1958]

$$\hat{c}_{\boldsymbol{k}\uparrow} = u_k \hat{\zeta}_{\boldsymbol{k}} + v_k \hat{\xi}^\dagger_{-\boldsymbol{k}}, \qquad \hat{c}_{-\boldsymbol{k}\downarrow} = -v_k \hat{\zeta}^\dagger_{\boldsymbol{k}} + u_k \hat{\xi}_{-\boldsymbol{k}} \qquad (4.36)$$

and Hermitian conjugate relations. The inverse transformation is

$$\hat{\zeta}_{\boldsymbol{k}} = u_k \hat{c}_{\boldsymbol{k}\uparrow} - v_k \hat{c}^\dagger_{-\boldsymbol{k}\downarrow}, \qquad \hat{\xi}_{-\boldsymbol{k}} = v_k \hat{c}^\dagger_{\boldsymbol{k}\uparrow} + u_k \hat{c}_{-\boldsymbol{k}\downarrow}. \qquad (4.37)$$

It is easily seen (exercise) that the transformation is canonical, if one chooses

$$u_k^2 + v_k^2 = 1, \qquad (u_k, v_k \text{ real}). \qquad (4.38)$$

The operator $\hat{\zeta}^\dagger_{\boldsymbol{k}}$ creates a $(\boldsymbol{k}\uparrow)$-particle and $\hat{\xi}^\dagger_{\boldsymbol{k}}$ creates a $(\boldsymbol{k}\downarrow)$-particle, both by hybridizing primary particle and hole excitations.

One easily computes (exercise)

$$(\hat{c}^\dagger_{\boldsymbol{k}\uparrow} \hat{c}_{\boldsymbol{k}\uparrow} + \hat{c}^\dagger_{-\boldsymbol{k}\downarrow} \hat{c}_{-\boldsymbol{k}\downarrow}) =$$
$$= 2v_k^2 + (u_k^2 - v_k^2)(\hat{\zeta}^\dagger_{\boldsymbol{k}} \hat{\zeta}_{\boldsymbol{k}} + \hat{\xi}^\dagger_{-\boldsymbol{k}} \hat{\xi}_{-\boldsymbol{k}}) + 2u_k v_k (\hat{\xi}_{-\boldsymbol{k}} \hat{\zeta}_{\boldsymbol{k}} + \hat{\zeta}^\dagger_{\boldsymbol{k}} \hat{\xi}^\dagger_{-\boldsymbol{k}}). \qquad (4.39)$$

The transformation of the interaction term of (4.35) and its subsequent rearrangement to normal order in the new operators is a somewhat tedious exercise, the result is explicitly only needed up to terms containing at most two operators. If the two abbreviations

$$\eta_k \stackrel{\text{def}}{=} \varepsilon^0_k - \mu - v_k^2 V_0, \qquad \Delta_k \stackrel{\text{def}}{=} \sum_{\boldsymbol{k}'} u_{k'} v_{k'} V_{|\boldsymbol{k}'-\boldsymbol{k}|} \qquad (4.40)$$

are introduced, the result for the transformed Hamiltonian is obtained as

$$\begin{aligned}\hat{H}_\mu &= \sum_{\boldsymbol{k}} (2v_k^2 \eta_k + v_k^4 V_0 - u_k v_k \Delta_k) + \\ &+ \sum_{\boldsymbol{k}} \left[(u_k^2 - v_k^2)\eta_k + 2u_k v_k \Delta_k\right] (\hat{\xi}^\dagger_{-\boldsymbol{k}} \hat{\xi}_{-\boldsymbol{k}} + \hat{\zeta}^\dagger_{\boldsymbol{k}} \hat{\zeta}_{\boldsymbol{k}}) + \\ &+ \sum_{\boldsymbol{k}} \left[2u_k v_k \eta_k - (u_k^2 - v_k^2)\Delta_k\right] (\hat{\zeta}^\dagger_{\boldsymbol{k}} \hat{\xi}^\dagger_{-\boldsymbol{k}} + \hat{\xi}_{-\boldsymbol{k}} \hat{\zeta}_{\boldsymbol{k}}) + \\ &+ \hat{H}'_{\text{int}}. \end{aligned} \qquad (4.41)$$

The not written out part $\hat{H}'_{\text{int}}$ consists of terms containing more than two operators as factors and describes remaining interactions. Since it is normal ordered, its expectation value in the vacuum of the $\hat{\zeta}$, $\hat{\xi}$-operators vanishes like all terms of this transformed Hamiltonian except the first line of (4.41). The anomalous contribution expressed by the third line can be made vanish as an operator, if one poses the constraint

$$(u_k^2 - v_k^2)\Delta_k = 2u_k v_k \eta_k \tag{4.42}$$

onto the transformation coefficients u_k and v_k of which only one was fixed by the canonicity condition (4.38).

One solution of the equations (4.38, 4.42) would be

$$u_k^2 = \theta(\eta_k), \qquad v_k^2 = \theta(-\eta_k) \tag{4.43}$$

and hence,

$$u_k v_k = 0 \implies \Delta_k = 0. \tag{4.44}$$

This brings one immediately back to (4.7), with only the interaction energy

$$\frac{1}{2}\sum_{k\sigma}^{\eta_k<0} V_0 \tag{4.45}$$

added to $\Omega(\mu)$. In order to find a non-trivial solution we rewrite the equations (4.38, 4.42) as

$$u_k = \cos\omega_k, \quad v_k = \sin\omega_k, \quad \Delta_k \cos 2\omega_k = \eta_k \sin 2\omega_k. \tag{4.46}$$

Defining

$$\tilde{\varepsilon}_k = \sqrt{\eta_k^2 + \Delta_k^2}, \tag{4.47}$$

the solution may be written as

$$\pm\frac{\eta_k}{\tilde{\varepsilon}_k} = \cos 2\omega_k = u_k^2 - v_k^2 = 2u_k^2 - 1 = 1 - 2v_k^2,$$

$$\pm\frac{\Delta_k}{\tilde{\varepsilon}_k} = \sin 2\omega_k = 2u_k v_k. \tag{4.48}$$

The signs must of course be chosen equal in both lines. Together with (4.40), the final result is

$$u_k^2 = \frac{1}{2}\left(1 + \frac{\eta_k}{\tilde{\varepsilon}_k}\right), \qquad v_k^2 = \frac{1}{2}\left(1 - \frac{\eta_k}{\tilde{\varepsilon}_k}\right), \tag{4.49}$$

$$\Delta_k = \frac{1}{2}\sum_{k'}\frac{\Delta_{k'}}{\tilde{\varepsilon}_{k'}}V_{|k'-k|}, \tag{4.50}$$

$$\hat{H}_\mu = \Omega(\mu) + \sum_k (\hat{\xi}_k^\dagger \tilde{\varepsilon}_k \hat{\xi}_k + \hat{\zeta}_k^\dagger \tilde{\varepsilon}_k \hat{\zeta}_k) + \hat{H}'_{\text{int}}. \tag{4.51}$$

In order that the second term of this Hamiltonian be positive definite, the positive signs have to be taken in (4.48). The condition for this solution to exist is that the gap equation (4.50) has a non-trivial solution.

The physical meaning of this superconducting solution is discussed in textbooks of superconductivity (e.g. [Schrieffer, 1964]). Our focus was to demonstrate at least approximately the transformation of a Hamiltonian of an interacting fermion system into a c-number grand canonical thermodynamic potential term plus a positive definite operator, which vanishes in the ground state with vanishing fluctuations (we considered a $T = 0$ theory), but which describes the dynamics of excitations with an energy dispersion law (4.47), in the present case exhibiting an energy gap Δ_{k_f}, where k_f is the Fermi momentum.

4.4 Weakly Interacting Bose Gas

Finally, we consider an isotropic Bose gas with short-range repulsive pair interaction and the Hamiltonian

$$\hat{H}_\mu = \sum_k \hat{b}_k^\dagger (\varepsilon_k^0 - \mu)\hat{b}_k + \sum_{kk'q} \hat{b}_{k+q}^\dagger \hat{b}_{k'-q}^\dagger \frac{g}{2V} \hat{b}_{k'}\hat{b}_k. \tag{4.52}$$

The free-particle energy is $\varepsilon_k^0 = \hbar^2 k^2/2m$, and the interaction matrix element $g/2V > 0$ is independent of the transmitted momentum q as a δ-like repulsion in real space. Since the sum over q is according to (1.55) proportional to the volume V, the constant matrix element of a volume-independent interaction strength must be inversely proportional to the volume.

We are again interested in the situation at zero temperature and expect Bose condensation. Hence, as a first step we single out the operators of the lowest energy state ($k = 0$), and put

$$\hat{b}_0^\dagger \to \sqrt{N_0}, \quad \hat{b}_0 \to \sqrt{N_0}. \tag{4.53}$$

The condensate density is denoted by $n_0 = N_0/V$. This first step transforms the

Hamiltonian into

$$\hat{H}_\mu = V\frac{gn_0^2}{2} + {\sum_{k}}' \hat{b}_k^\dagger(\varepsilon_k^0 - \mu + 2gn_0)\hat{b}_k +$$
$$+ \frac{gn_0}{2} {\sum_{k}}' (\hat{b}_k^\dagger \hat{b}_{-k}^\dagger + \hat{b}_k \hat{b}_{-k}) + \hat{H}'_{\text{int}}. \tag{4.54}$$

$\hat{H}'_{\text{int}}$ consists of terms with products of at least three operators. The prime at the sums means leaving out the item with $\bm{k} = 0$. The four contributions with products of two $\hat{b}$-operators arise from the cases $\bm{k}' = \bm{q} = 0$, $\bm{k} = \bm{q} = 0$, $\bm{k}' = \bm{k} + \bm{q} = 0$, and $\bm{k} = \bm{k}' - \bm{q} = 0$ in (4.52).

Next we perform a canonical transformation, first applied to a boson-like spin situation by [Holstein and Primakoff, 1940] and later to the present case by [Bogolubov, 1947]:

$$\hat{b}_k = u_k \hat{\beta}_k - v_k \hat{\beta}_{-k}^\dagger, \quad u_k^2 - v_k^2 = 1, \quad u_k, v_k \text{ real} \tag{4.55}$$

and the Hermitian conjugate relation for $\hat{b}_k^\dagger$. The canonicity is easily verified (exercise). This transforms the Hamiltonian into (exercise)

$$\hat{H}_\mu = V\frac{gn_0^2}{2} + {\sum_{k}}' \left[v_k^2(\varepsilon_k^0 - \mu + 2gn_0) - u_k v_k gn_0\right] +$$
$$+ {\sum_{k}}' \hat{\beta}_k^\dagger \left[(u_k^2 + v_k^2)(\varepsilon_k^0 - \mu + 2gn_0) - 2u_k v_k gn_0\right] \hat{\beta}_k +$$
$$+ {\sum_{k}}' \left[\frac{gn_0}{2}(u_k^2 + v_k^2) - u_k v_k(\varepsilon_k^0 - \mu + 2gn_0)\right] *$$
$$* (\hat{\beta}_k^\dagger \hat{\beta}_{-k}^\dagger + \hat{\beta}_k \hat{\beta}_{-k}) +$$
$$+ \hat{H}'_{\text{int}}. \tag{4.56}$$

Again, the anomalous terms of the third line can be made vanish by a constraint to u_k and v_k, namely

$$gn_0(u_k^2 + v_k^2) = 2(\varepsilon_k^0 - \mu + 2gn_0)u_k v_k. \tag{4.57}$$

This time we may incorporate the canonicity constraint by putting

$$u_k = \cosh \omega_k, \quad v_k = \sinh \omega_k \tag{4.58}$$

4.4 Weakly Interacting Bose Gas

and have

$$gn_0 \cosh 2\omega_k = (\varepsilon_k^0 - \mu + 2gn_0) \sinh 2\omega_k, \tag{4.59}$$

which is solved with

$$\tilde{\varepsilon}_k = \sqrt{(\varepsilon_k^0 - \mu + 2gn_0)^2 - (gn_0)^2} \tag{4.60}$$

by

$$\frac{gn_0}{\tilde{\varepsilon}_k} = \sinh 2\omega_k = 2u_k v_k,$$

$$\frac{\varepsilon_k^0 - \mu + 2gn_0}{\tilde{\varepsilon}_k} = \cosh 2\omega_k = u_k^2 + v_k^2 = 2u_k^2 - 1 = 1 + 2v_k^2. \tag{4.61}$$

The final result is

$$u_k^2 = \frac{1}{2}\left(\frac{\varepsilon_k^0 - \mu + 2gn_0}{\tilde{\varepsilon}_k} + 1\right), \quad v_k^2 = \frac{1}{2}\left(\frac{\varepsilon_k^0 - \mu + 2gn_0}{\tilde{\varepsilon}_k} - 1\right), \tag{4.62}$$

$$\hat{H}_\mu = \frac{1}{2}\left(Vgn_0^2 - \sum_{\boldsymbol{k}}{}'[\varepsilon_k^0 - \mu + 2gn_0 - \tilde{\varepsilon}_k]\right) + \sum_{\boldsymbol{k}}{}'\hat{\beta}_{\boldsymbol{k}}^\dagger \tilde{\varepsilon}_k \hat{\beta}_{\boldsymbol{k}} + \hat{H}'_{\text{int}} =$$

$$= \Omega(\mu) + \sum_{\boldsymbol{k}}{}'\hat{\beta}_{\boldsymbol{k}}^\dagger \tilde{\varepsilon}_k \hat{\beta}_{\boldsymbol{k}} + \hat{H}'_{\text{int}}. \tag{4.63}$$

The first, c-number term is again the grand canonical thermodynamic potential and the second is a positive definite operator yielding zero energy with zero fluctuations (we again considered $T = 0$) when applied to the $\hat{\beta}$-vacuum.

In the first term of (4.63), the $\boldsymbol{k}$-sum is to be cut at some $k_0 \sim r_0^{-1}$, where r_0 is the radius of the repulsive particle core. For larger $\boldsymbol{k}$, the interaction strength $g \to 0$ and hence $\tilde{\varepsilon}_k \to \varepsilon_k^0 - \mu$. If the condensate term is dominating, $N_0 \approx N$, then $\Omega \approx gN^2/2V$ and hence $\mu = \partial\Omega/\partial N \approx gn \approx gn_0$, $\varepsilon_k^0 - \mu + 2gn_0 \approx \varepsilon_k^0 + gn_0$. Although the primary particles had a quadratic energy dispersion relation, the energy dispersion $\tilde{\varepsilon}_k$ for the elementary excitations is *linear* in this case,

$$\tilde{\varepsilon}_k \approx \left(\frac{gn_0}{m}\right)^{1/2} \hbar k \quad \text{for} \quad k \to 0. \tag{4.64}$$

A more subtle consideration shows that this is indeed correct and gives the correct velocity of sound in the Bose liquid.

4.5 Discussion of the Model Approach

The general approach of this chapter can be characterized as follows: We are given some Hamiltonian

$$\hat{H}_\mu = \sum_\alpha \hat{a}_\alpha^\dagger (\varepsilon_\alpha^0 - \mu) \hat{a}_\alpha + \hat{H}_{\text{int}} \qquad (4.65)$$

and some 'nonempty' spatially homogeneous state $|\Psi_0\rangle$,

$$\hat{a}_\alpha |\Psi_0\rangle \neq 0. \qquad (4.66)$$

The aim is to cast this Hamiltonian into a form (the number of letters is painfully finite; $\hat{b}$ means here a general annihilator, not necessarily bosonic)

$$\hat{H}_\mu = \Omega(T, V, \mu) + \sum_\beta \hat{b}_\beta^\dagger (\tilde{\varepsilon}_\beta - \mu) \hat{b}_\beta + \hat{H}'_{\text{int}}, \qquad (4.67)$$

so that the former state $|\Psi_0\rangle$ is now the vacuum of the $\hat{b}$-operators,

$$|\Psi_0\rangle \hat{=} |\rangle, \qquad \hat{b}_\beta |\rangle = 0 \text{ for all } \beta. \qquad (4.68)$$

Here, $\Omega(T, V, \mu)$ does not influence the local dynamics, in particular not the propagation of elementary excitations. It is a c-number on each *irreducible* sector of the representation space of the $\hat{b}$-operators. However, as a thermodynamic potential, it governs the thermodynamics.

The propagation of elementary excitations above the state $|\Psi_0\rangle$ is governed by the model Hamiltonian

$$\hat{H}_{\text{mod}} = \sum_\beta \hat{b}_\beta^\dagger (\tilde{\varepsilon}_\beta - \mu) \hat{b}_\beta + \hat{H}'_{\text{int}}, \qquad (4.69)$$

for which the state $|\Psi_0\rangle$ is the vacuum (of elementary excitations), and wherein the interaction term $\hat{H}'_{\text{int}}$ hopefully is *weak*, so that $\tilde{\varepsilon}_\beta$ approximates the excitation energy of an elementary excitation, generally given by (3.33):

$$\tilde{\varepsilon}_\beta \approx \text{Re } \varepsilon_\beta. \qquad (4.70)$$

Since the transformation from $\hat{H}_\mu$ to $\hat{H}_{\text{mod}}$ depends on the thermodynamic state $|\Psi_0\rangle$ (in the simplest case, at $T = 0$, it depends on the chemical potential μ and hence on the average density), the quantities entering the model Hamiltonian, $\tilde{\varepsilon}_\beta$,

4.5 Discussion of the Model Approach

the quantum numbers β themselves, and the interaction matrix elements, generally also depend on the thermodynamic variables. Changing for instance the density or the temperature or the pressure means replacing $|\rangle$ again by some $|\Psi_0\rangle$, and a new contribution to Ω arises together with a change of the model parameters. For a finite system in a box, sufficiently mild changes can be treated within one model representation. Generally, however, a model is only useful in a limited range of situations, because the influence of the interaction term on the dynamics may become indescribably chaotic if the $\hat{b}_\beta^\dagger$ do not create a relevant part of the elementary excitations, so that the pole term of the corresponding Green's function (3.32) becomes negligibly small.[29]

Only in simple situations as for instance those of the last two sections, the transformation to the model Hamiltonian may be carried through explicitly. In most cases, if it can be derived at all by means of theory, uncontrolled approximations are necessary (even though often with astonishing success). Often one simply has to have resort to phenomenology.

In order to describe, say, a metal, the Hamiltonian of Bloch electrons and phonons must be given. However, now the excited state of a single Bloch electron is unstable. It looses energy by emitting phonons. Hence, the numbers which should appear in the diagonal part of the Hamiltonian are only approximately equal to the experimentally observed quasi-particle energies ε_β (experimentally seen is a renormalized energy). The same is to be said about the gas part of the partition function (3.35), and just this is the reason why in (3.36) a sign of approximate equality appeared. The situation is exactly as familiar in quantum field theory of particle physics, where the *bare* masses of the Lagrangian are not the observed ones. As will be seen, it is also not possible to rigorously derive the needed Hamiltonian from the quantum chemical one of nuclei and electrons interacting with Coulomb forces (to say nothing about the approximate character of this Hamiltonian in the frame of Quantum Electrodynamics). Only in recent years the simplest interaction terms of the Hamiltonian in solid state theory were calculated 'from the first principles': the anharmonicity in semiconductors [Deinzer *et al.*, 2004, and references therein], the non-adiabatic electron-phonon interaction [Kouba *et al.*, 2001, Doğan and Marsiglio, 2003, Deppeler and Millis, 2002] and the on site Coulomb matrix element of the Hubbard model [Dederichs *et al.*, 1984, Norman and Freeman, 1986, Gunnarsson *et al.*, 1989]. All that shows that the quasi-particle Hamiltonian has always a model character.

---

[29] In the case of a fermionic model it has become custom to speak of non-Fermi-liquid behavior in that situation.

Of course, constructing it one starts from a physical picture. So, in the case of a metal, one supposes the existence of charged electrons, interacting with screened Coulomb forces (the screened interaction matrix elements entering the Hamiltonian are, however, not exactly known, and not even uniquely determined), the existence of collective motions of the nuclei described by phonons, and so on.

Precisely on such a model Hamiltonian the elementary electronic theory of metals by A. Sommerfeld was based [Ashcroft and Mermin, 1976, Chapter 2], and its success historically gave the first indication of the correctness of the type of theory sketched here, though the theoretical reasoning along the lines of the last sections appeared much later. Another (analogous) example is the Debye theory of thermal properties of solids, based on the model Hamiltonian of phonons [Ashcroft and Mermin, 1976, Chapter 23].

At this point, some remarks are in due place on the model (4.69). First of all, the possibility to describe the solid at low excitation energies *completely* by the Hamiltonian (4.69) is an experience of solid state theory. We have at hand theoretical means to establish individual branches of excitations of a given concrete character, the nature of these excitations itself, however, is *guessed* with the help of physical intuition, often using experimental facts. This situation does not simply demonstrate the imperfection of our present theoretical abilities, but has a deeper origin in the principally approximative character and, as will be discussed below, ambiguity of the whole approach. Particularly it seems not to be possible to formally *demonstrate*, under which general conditions the quasi-stationary excitations exhaust the whole low-energy dynamics of the macrosystem. For instance, little is known about the electronic subsystem of an amorphous solid at very low energies [Mott and Davis, 1979, Lifshits *et al.*, 1988, Shklovskii and Efros, 1984, Ortuño *et al.*, 2001]; here it is especially difficult to find arguments in favor of the condition (3.34).

Furthermore, the aim of constructing a Hamiltonian of the type (4.69) is anything but to diagonalize the 'original' electron-ion Hamiltonian as a Hilbert space operator. Such a diagonalization would just deliver the *stationary* levels E_i of Section 2.4 the uselessness of which was already discussed in that section. Instead, the model (4.69) is built so that its Green's functions would have the same low-energy poles (3.33) as the Green's functions of the original system (in the approximative sense of the above discussed realistic assumption), but this time they may be calculated using perturbation theory, resulting in a renormalization $\delta\tilde{\varepsilon}_\beta = \mathrm{Re}\,\varepsilon_\beta - \tilde{\varepsilon}_\beta$ of the energy and in a damping $|\mathrm{Im}\,\varepsilon_\beta| > 0$. If one speaks of 'diagonalizing the Hamiltonian' in this context in the literature one has in mind to transform it in

4.5 Discussion of the Model Approach

such a way that the first part of (4.69) is more strongly dominating over the second. Also within *this* frame it is impossible in any way to transform the Hamiltonian (4.69) so as to include $\hat{H}_{\text{int}}$ into the diagonal and to obtain $\hat{H}'_{\text{mod}} = \sum_\beta \hat{b}^\dagger_\beta \varepsilon_\beta \hat{b}_\beta$. Without mentioning that $\hat{H}'_{\text{mod}}$ would not be Hermitian (the ε_α are complex), the appearance of the term $\hat{H}_{\text{int}}$ in (4.69) has at least two physical consequences: first the renormalization and damping of the energies depend on the state of the system (see below), especially in the thermodynamic equilibrium they depend on temperature, second $\hat{H}_{\text{int}}$ describes the creation of particles of one sort during the decay of others being particularly important in kinetics.

5 Quasi-Particles

In the important special case when the set α of parameters contains the momentum $\boldsymbol{p}$ of the excitation of a translationally invariant system or the quasi-momentum $\boldsymbol{p}$ mod $\hbar\boldsymbol{G}$ in a crystalline lattice with $\boldsymbol{G}$ as a vector in the reciprocal lattice, i.e., (ν denotes the remaining parameters)

$$\varepsilon_\alpha \equiv \varepsilon_\nu(\boldsymbol{p}), \tag{5.1}$$

one speaks of *quasi-particles* and their *dispersion relation* (5.1). In this terminology the state Ψ_0 of (3.14) may be characterized as the state without quasi-particles or as the *quasi-particle vacuum*. This appears natural in the representations (4.6) and (4.15) where this quasi-particle vacuum is symbolized by $|\rangle$. In the fermion case both particles and holes are likewise called quasi-particle in this context.

As the temperature is gradually raised, quasi-particles are created, but their number is in the beginning rather small and, consequently, their mean distance large and their interaction weak. Hence, the solid may be treated as a *dilute gas of weakly interacting quasi-particles with an arbitrary dispersion law*[30]. Comparing a gas of quasi-particles to a gas of ordinary particles, emphasis lies on the involved character of the quasi-particle dispersion law (5.1); for a crystal, $\varepsilon_\nu(\boldsymbol{p})$ is a periodic function of $\boldsymbol{p}$ with the periods $\hbar\boldsymbol{G}$. The determination of this dispersion law is one of the basic tasks of solid state spectroscopy.

Up to now we have preferentially used the more flexible term 'quasi-stationary excitation' comprising also situations where the set of parameters α does not contain the momentum, e.g. impurity levels in a semiconductor or localized states in disordered systems. In this chapter we focus on the dispersion relation and use the name quasi-particle in this narrower sense. In this context, a well defined dispersion relation, that is only a discrete set of energies attached with each momentum $\boldsymbol{p}$, distinguishes single-particle excitations, which are elementary excitations,

---

[30]In the important case of phonons the quasi-particle interaction is small due to the large masses of nuclei compared to that of electrons responsible for the bonding forces. For this reason, the phonon gas is at any temperature (up to near the melting point) weakly non-ideal with clearly observed individual phonons [Eckert and Youngblood, 1986, Horton and Cowley, 1987]. Eventual exceptions are the temperature intervals around second order structural phase transitions (see for instance [Labbe and Friedel, 1966a, b, c]).

from multi-particle excitations, which are not elementary. For them, continuous intervals of energy are available at each momentum. Compare also the previous discussion in connection with (3.35). In the literature the terms 'quasi-particle' and 'quasi-stationary excitation' are often used as synonyms.

5.1 Landau's Quasi-Particle Conception

The quasi-particle conception for the description of the excitation spectrum of condensed matter in the sense of (3.35) was first introduced by L. D. Landau [Landau, 1941] and successfully applied to explain superfluidity and the thermodynamic properties of the superfluid liquid ^{4}He. Later on [Landau, 1956], he applied this notion to describe the properties of the normal (i.e. not superfluid) Fermi liquid ^{3}He. B. M. Galitzkii and A. B. Migdal [Galitskii and Migdal, 1958] derived the consistent microscopic theory of quasi-particles using the apparatus of Green's functions. The quasi-particle notion has since proved to be fundamental for the whole of condensed matter theory (see also [Kaganov and Liftshits, 1980]).

One could think that, if there is only one quasi-particle above the vacuum, there would be nothing to interact with and it should be stable. However, a physical vacuum can be polarized by exciting particle-antiparticle pairs, in a solid for instance by exciting electron-hole pairs, or by exciting neutral particles, as for instance phonons. Only if the single present particle has a low excitation energy, the number of accessible vacuum polarization channels is limited by energy conservation and the particle can be long-lived. If its excitation energy tends to zero, its lifetime may tend to infinity.

Thus, as long as the number of quasi-particles is sufficiently small and their mutual interaction is weak, the state of the system may be well characterized by the mean occupation numbers n_α of the quasi-particle states. The total energy of the system may then be expressed as a functional $E[n_\alpha]$ of these numbers, and

$$\xi_\alpha = \frac{\delta E}{\delta n_\alpha} \tag{5.2}$$

may be taken as the (real) quasi-particle energy (cf. (3.36)).

The quasi-particle interaction leads in this approach to a non-linear dependence of the energy functional on the n_α and hence to a *dependence of the quasi-particle energy ξ_α on the averaged number of quasi-particles present, i.e., on the state of* *the system* (see the remarks at the end of the last chapter):

$$\frac{\delta(\xi_\alpha)}{\delta n_\beta} = \frac{\delta^2 E}{\delta n_\beta \delta n_\alpha}, \quad \delta(\xi_\alpha) = \int d\beta \, \delta n_\beta \, \frac{\delta^2 E}{\delta n_\beta \delta n_\alpha}. \tag{5.3}$$

Precisely on these formulas (5.2) and (5.3) the phenomenological Landau theory [Landau, 1956] was based; see also [Landau and Lifshits, 1980b].

Let the second functional derivative in (5.3) be denoted by

$$f(\alpha, \beta) = \frac{\delta^2 E}{\delta n_\alpha \delta n_\beta} = f(\beta, \alpha). \tag{5.4}$$

Obviously it is symmetric in its arguments by definition. Integration of the second equation (5.3) yields

$$\xi_\alpha[n_\beta] = \xi_\alpha^0 + \int d\beta \int_0^{n_\beta} dn_\beta \, f(\alpha, \beta). \tag{5.5}$$

the first term, ξ_α^0 is the energy of a singly excited quasi-particle α, and the second term is the change in its excitation energy due to interaction with all other quasi-particles β, on the average present with occupation numbers n_β. Therefore Landau called $f(\alpha, \beta)$ the quasi-particle interaction function. (Strictly speaking, $f(\alpha, \beta)[n_\gamma]$ is again a functional of the occupation numbers n_γ. Since the theory is only meant for small occupation numbers as discussed in the introduction to this chapter, this latter functional dependence is neglected.)

Consider the effect of the interaction function $f(\alpha, \beta)$ on transport properties [Abrikosov, 1988]. In the quasi-classical linear transport theory the kinetic equation

$$\frac{\partial n_\alpha}{\partial t} + \frac{\partial n_\alpha}{\partial \bm{r}} \frac{d\bm{r}}{dt} + \frac{\partial n_\alpha}{\partial \bm{p}} \frac{d\bm{p}}{dt} = \left[\frac{dn_\alpha}{dt}\right]_{\text{coll.}} \tag{5.6}$$

for the occupation numbers is considered the right hand side of which is the collision integral. To be specific, consider electrons (with charge $e < 0$ in an external electric field $\bm{E}$. The force of the field on the electron is $d\bm{p}/dt = e\bm{E}$. If, however, $\partial n_\beta / \partial \bm{r} \neq 0$ results, then via (5.3) there is an additional force of a self-consistent field, given by $-\partial \xi_\alpha / \partial \bm{r}$:

$$\frac{d\bm{p}}{dt} = e\bm{E} - \int d\beta \, \frac{\partial n_\beta}{\partial \bm{r}} \, f(\alpha, \beta). \tag{5.7}$$

5.1 Landau's Quasi-Particle Conception

The linear theory is obtained by putting

$$n_\alpha = n_0(\xi_\alpha[n_0]) - \psi_\alpha \frac{dn_0}{d\xi_\alpha} \tag{5.8}$$

and linearizing in the deviation factor ψ from the thermal equilibrium occupation $n_0(\xi[n_0])$. With $\partial n_0/\partial \boldsymbol{p} = (dn_0/d\xi)(\partial \xi/\partial \boldsymbol{p}) = (dn_0/d\xi)\,\boldsymbol{v}$ and after dividing the linear terms by $dn_0/d\xi$ one finds

$$\frac{\partial \psi_\alpha}{\partial t} + \boldsymbol{v}_\alpha \cdot \frac{\partial}{\partial \boldsymbol{r}} \left(\psi_\alpha - \int d\beta \, \frac{dn_0}{d\xi_\beta} \, f(\alpha,\beta)\, \psi_\beta \right) - e\boldsymbol{v}_\alpha \cdot \boldsymbol{E} = \left[\frac{d\psi_\alpha}{dt}\right]_{\text{coll.}} \tag{5.9}$$

The collision integral is zero for $n = n_0(\xi)$ for the true particles present (with energy ξ). Combining (5.8) and (5.5) into

$$n_\alpha = n_0(\xi_\alpha[n_\alpha]) - \frac{dn_0}{d\xi_\alpha}\left[\psi_\alpha - \int d\beta\, \frac{dn_0}{d\xi_\beta}\, f(\alpha,\beta)\, \psi_\beta\right], \tag{5.10}$$

and realizing that the collision integral is zero for the first term, it is clear that the linearized collision integral must depend on

$$\varphi_\alpha = \psi_\alpha - \int d\beta\, \frac{dn_0}{d\xi_\beta}\, f(\alpha,\beta)\, \psi_\beta\,. \tag{5.11}$$

Finally, the electric current density is

$$\begin{aligned}
\boldsymbol{j} &= e\int d\alpha\, \frac{\partial \xi_\alpha}{\partial \boldsymbol{p}}\, n_\alpha = e\int d\alpha\, \left(\frac{\partial \xi_\alpha[n_0]}{\partial \boldsymbol{p}}\, \delta n_\alpha + \frac{\partial \delta \xi_\alpha}{\partial \boldsymbol{p}}\, n_0(\xi_\alpha[n_0])\right) = \\
&= e\int d\alpha\, \left(\boldsymbol{v}_\alpha \delta n_\alpha + n_0(\xi_\alpha[n_0])\, \frac{\partial}{\partial \boldsymbol{p}}\int d\beta\, f(\alpha,\beta)\, \delta n_\beta\right) = \\
&= e\int d\alpha\, \left(\boldsymbol{v}_\alpha \delta n_\alpha - \frac{dn_0}{d\xi_\alpha}\, \boldsymbol{v}_\alpha \int d\beta\, f(\alpha,\beta)\, \delta n_\beta\right) = \\
&= -e\int d\alpha\, \frac{dn_0}{d\xi_\alpha}\, \boldsymbol{v}_\alpha \left(\psi_\alpha - \int d\beta\, \frac{dn_0}{d\xi_\beta}\, f(\alpha,\beta)\, \psi_\beta\right).
\end{aligned} \tag{5.12}$$

In the third line an integration per parts was performed, and then again $\delta n = -\psi dn_0/d\xi$ was inserted.

It is now readily seen that in stationary transport, when $\partial \psi/\partial t \approx 0$, the whole kinetic theory contains $f(\alpha,\beta)$ only in the function φ_α of (5.11) which is, however, the solution of the kinetic equation (5.9), otherwise not containing $f(\alpha,\beta)$. The same can be shown for any transport quantity. *The stationary transport does not depend on the interaction function $f(\alpha,\beta)$.*

The resistivity $\sim T^2$ from electron-electron scattering for instance arises from scattering processes with formation of new electron-hole pairs or their mutual annihilation, that is from processes involving vacuum polarization.

5.2 Quasi-Particles in a Crystalline Solid

As was already mentioned at the end of the last chapter, the model of free non-interacting electrons,

$$\varepsilon(\boldsymbol{p}) = p^2/2m, \tag{5.13}$$

was used by A. Sommerfeld in his elementary theory of metals immediately after the discovery of quantum mechanics. (On the basis of classical mechanics, that is without taking account of the Pauli principle, the corresponding model was introduced by P. Drude already at the beginning of last century.) Taking into consideration the interaction of the electrons with the periodic crystal field, F. Bloch founded the band model of non-interacting with each other electrons

$$\varepsilon_\alpha = \varepsilon_\nu(\boldsymbol{p} \bmod \hbar \boldsymbol{G}), \tag{5.14}$$

where ν is the band (and spin) index. Despite its simplicity, this model not only explains why crystalline solids appear either to be metals or insulators, semiconductors or semi-metals, it also explains many of the special electronic properties of solids: the valences of transition metals, at least semi-quantitatively the value of the heat capacity of metals, the appearance of both signs of the Hall coefficient and of the thermopower, and so on. True, the 'explanation' has often been based on reasoning as: 'if the dispersion law is this and this, then the properties are those $\cdots$' Thereby the dispersion law has been chosen among those principally admissible for the motion of an electron in a lattice-periodic potential of a certain symmetry. Even nowadays, when quasi-particle band structures are calculated with great reliability in many cases, there remain whole classes of solids for which one has to resort to the above reasoning.

Already on the basis of modern understanding of Bloch electrons as weakly interacting quasi-particles with the dispersion law (5.14), I. M. Lifshits and his coworkers put (and to a considerably extent solved) the 'inverse problem' to extract from the various experimental data the actual dispersion law (5.14) for

5.2 Quasi-Particles in a Crystalline Solid

metals at low excitation energies (mainly from the behavior in magnetic field [Lifshits et al., 1973]). Since the quasi-momenta of conduction electrons with low excitation energies lie in the vicinity of the *Fermi surface*

$$\varepsilon_\nu(\boldsymbol{p}_F^{(\nu)} \bmod \hbar \boldsymbol{G}) = \varepsilon_F \equiv \mu(T=0), \tag{5.15}$$

this Fermi surface and the velocity distribution on it, $\boldsymbol{v}_F^{(\nu)} = \partial \varepsilon_\nu / \partial \boldsymbol{p}|_{\boldsymbol{p}_F^{(\nu)}}$, are to be determined ('monster with beard'). An analogous program for semiconductors (mainly using their optical properties) was essentially performed by M. L. Cohen and T. K. Bergstresser by fitting to the experimental data an *empirical crystalline pseudopotential* from which the dispersion relation (5.14) is obtained. See [Cohen and Heine, 1970] for a more historical overview over applications of empirical pseudopotentials to both semiconductors and metals.

In quantum theory the momentum $\boldsymbol{p}$ of a particle becomes a measurable quantity due to spatial translational symmetry: since a momentum eigenstate is an extended plane wave, it needs a large homogeneous volume to precisely measure it. In a crystal with basis vectors $\boldsymbol{a}_i$, $i = 1, 2, 3$ there is only discrete translational symmetry with respect to lattice translations $\boldsymbol{R}_n = \sum_{i=1}^{3} n_i \boldsymbol{a}_i$. A wave function state of a particle covariant under this symmetry is a Bloch state $\phi(\boldsymbol{r} + \boldsymbol{R}_n) = \exp(i\boldsymbol{p} \cdot \boldsymbol{R}_n / \hbar)\phi(\boldsymbol{r})$ where now $\boldsymbol{p}$ with

$$\boldsymbol{p} = \boldsymbol{p} \bmod \hbar \boldsymbol{G}_m, \quad \boldsymbol{G}_m \cdot \boldsymbol{R}_n = 2\pi \cdot \text{integer}, \tag{5.16}$$

is a quasi-momentum. Obviously, $\boldsymbol{G}_m$ is a reciprocal lattice vector: $\boldsymbol{G}_m = \sum_{i=1}^{3} m_i \boldsymbol{b}_i$, $\boldsymbol{b}_i \cdot \boldsymbol{a}_j = 2\pi \delta_{ij}$. A change of $\boldsymbol{p}$ by $\hbar \boldsymbol{G}_m$ means a Bragg scattering on the lattice. The quasi-momentum is uniquely determined only within the first Brillouin zone, which means $\boldsymbol{p} \cdot \boldsymbol{G}_m < \hbar |\boldsymbol{G}_m|^2 / 2$ for all $\boldsymbol{G}_m$. Any function of $\boldsymbol{p}$ is periodic in the reciprocal lattice like (5.14). (See Fig. 5 on the next page; here, the first Brillouin zone is the interval $-\pi/a \leq p/\hbar \leq \pi/a$.)

Like with plane waves, wave pockets may be formed with Bloch waves, and their group velocity will be likewise obtained from elementary wave theory to be

$$\boldsymbol{v}_\nu = \frac{\partial \varepsilon_\nu}{\partial \boldsymbol{p}}, \tag{5.17}$$

which is again a periodic function in the reciprocal lattice for each band ν.

Consider a Bloch electron state with energy $\varepsilon_\nu(\boldsymbol{p})$ corresponding to the dot in Fig. 5. It has a momentum vector $\boldsymbol{p}$ pointing to the right, but a wave pocket formed of wave functions out of that band ν and with momenta in the vicinity of

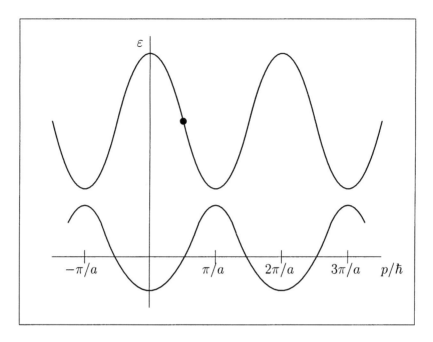

Figure 5: One-dimensional band structure $\varepsilon_\nu(p)$.

that momentum p is traveling to the left with group velocity v_ν. *The angle between the momentum vector and the velocity vector of a Bloch electron (or hole) can be quite arbitrary.* This is the main difference between a free particle and a particle moving in a lattice periodic potential.

If a force F acts on a particle, in (non-relativistic) quantum theory the operator equation $\dot{\hat{p}} = (i/\hbar)[\hat{H}, \hat{p}]_- = \hat{F}$ is valid (cf. (1.8)). For a wave pocket, the corresponding Ehrenfest relation $\dot{p} = F$ for the expectation values is valid. For a free particle this leads to the same relation

$$a = F/m \tag{5.18}$$

between the acceleration $a = \dot{v}$ and the force as Newton's law in classical mechanics. For a Bloch electron, (5.17) results in $\dot{v}_\nu = (\dot{p} \cdot \partial/\partial p)(\partial \varepsilon_\nu / \partial p)$ and hence in

$$a_j = \sum_i F_i (m_\nu^{-1})_{ij}, \qquad (m_\nu^{-1})_{ij} = \frac{\partial^2 \varepsilon_\nu}{\partial p_i \partial p_j}. \tag{5.19}$$

$m_\nu^{-1}(p)$ is the *inverse mass tensor* of the quasi-particle with the dispersion relation (5.17). Again, *the angle between the vector of an applied force and the acceleration vector of a Bloch particle can be quite arbitrary.* As a symmetric tensor, m_ν^{-1} can be diagonalized at any p. Its eigenvalues may have either sign. In Fig. 5 at the indicated dot, the quasi-particle mass just is changing sign from positive at lower energies to negative at higher energies.[31]

One more remark is in due place here. Considering the quasi-classical motion of a Bloch electron in an applied magnetic field, the electron may orbit clockwise or counterclockwise on the Fermi surface around the field direction as axis, depending on the topology of the Fermi surface. This also determines the sign of the Hall voltage. If the Fermi level is close above the bottom of a band, then the Fermi surface is closed and encloses occupied electron states. Then the inverse mass, the orbiting, the Hall voltage and so on is like for free electrons, and one speaks of *electron transport* and of an *electron Fermi surface*. If the Fermi level is close below the top of a band, then this is all reversed and one speaks of *hole transport* and of a *hole Fermi surface*. This terminology, which is also common in semiconductor physics, is related to that used in the present text, but is not the same.

5.3 Green's Function of Bloch Electrons

Here, $\varepsilon_\nu(p)$ is now understood as the solution of (3.33). For the sake of simplification of the analysis we drop spin dependences. They are commonly present and simply lead to a spin matrix character of the quantities considered but not to new aspects in the forthcoming considerations. Moreover, as electrons are considered, in most cases it suffices to treat them at $T = 0$, since on the electronic energy scale temperature is normally very low. G in the present case means the matrix $G_{GG'}$ defined by

$$\delta(p - p') \, G_{GG'}(\omega, p) =$$
$$\stackrel{\text{def}}{=} \int d^3r d^3r' \, e^{i(p+G)r/\hbar} \, G(\omega, r, r') \, e^{-i(p'+G')r'/\hbar} \qquad (5.20)$$

---

[31] The various effective masses introduced in Solid State Physics are all related to (5.19): The effective cyclotron mass determining the cyclotron frequency of a Bloch electron in an applied magnetic field is $m^* = \overline{((\det m_\nu)(m_\nu^{-1})_{zz})^{1/2}}$, where z is the direction of the magnetic field, and the average is over the cyclotron orbit of the Bloch electron. The effective mass related to the heat capacity is $m^*(\varepsilon) = \overline{\sum_\nu (\det m_\nu)^{1/3}}$, averaged over all p with $\varepsilon_\nu(p) = \varepsilon$. m_ν is the inverse of the tensor m_ν^{-1}. See, [Ashcroft and Mermin, 1976, Chapter 12].

with $G(\omega, \boldsymbol{r}, \boldsymbol{r}')$ defined as

$$G(\omega, \boldsymbol{r}, \boldsymbol{r}') = \int_{-\infty}^{\infty} dt \, e^{i\omega t/\hbar} G(t, \boldsymbol{r}, \boldsymbol{r}') \tag{5.21}$$

like in (3.20) and

$$G(t, \boldsymbol{r}, \boldsymbol{r}') = -i\langle|\mathbf{T}\hat{\psi}(\boldsymbol{r}, t)\hat{\psi}^\dagger(\boldsymbol{r}', 0)|\rangle, \tag{5.22}$$

where $|\rangle$ is the vacuum (4.6) and $\hat{\psi}(\boldsymbol{r}, t)$ is the Heisenberg operator corresponding to (4.10) or, which amounts to the same, $|\rangle$ is the ground state $|\Psi_0\rangle$ introduced in (3.14) and $\hat{\psi}(\boldsymbol{r}, t)$ is the Heisenberg operator corresponding to $\hat{\psi}(\boldsymbol{r}) = \sum_\alpha \phi_\alpha(\boldsymbol{r}) \hat{c}_\alpha$, $\alpha = (\nu, \boldsymbol{p})$ with ν also counting for the spin. The pole position of this Green's function has a positive real part for electrons and a negative real part for holes like in the right part of Fig. 4, likewise for the quasi-particle density of states, where in the quasi-particle spectral weight (3.30) G^r is now to be replaced with ZG^r. The quasi-particle spectral function Z was introduced in (3.32). (As in all cases, the sign of the imaginary part of the pole position is opposite to that of the real part.)

$G(\omega, \boldsymbol{r}, \boldsymbol{r}')$ obeys the Dyson equation [Landau and Lifshits, 1980b]

$$\left[\omega + \mu + \frac{\hbar^2 \nabla^2}{2m} - v_H(\boldsymbol{r})\right] G(\omega, \boldsymbol{r}, \boldsymbol{r}') -$$

$$- \int d^3 r'' \, \Sigma(\omega, \boldsymbol{r}, \boldsymbol{r}'') \, G(\omega, \boldsymbol{r}'', \boldsymbol{r}') = \hbar \delta(\boldsymbol{r} - \boldsymbol{r}'). \tag{5.23}$$

In this equation the Hartree potential v_H is as previously the total electrostatic potential, created by the charge density of all electrons and of fixed (on regular lattice sites) nuclei, and multiplied by the electron charge. The so-called self-energy operator Σ represents a specially selected partial sum of the perturbation series for G.

In the vicinity of the solution of (3.33) in the complex ω-plane, G takes on the form (3.32):

$$G_{\boldsymbol{G}\boldsymbol{G}'}(\omega, \boldsymbol{p}) = \sum_\nu \frac{\hbar Z^{(\nu)}_{\boldsymbol{G}\boldsymbol{G}'}(\boldsymbol{p})}{\omega + \mu - \varepsilon_\nu(\boldsymbol{p})} + g_{\boldsymbol{G}\boldsymbol{G}'}(\omega, \boldsymbol{p}), \tag{5.24}$$

5.3 Green's Function of Bloch Electrons

where $g_{GG'}$ has already no singularity for $\omega \to \varepsilon_\nu(\boldsymbol{p}) - \mu$. Substituting (5.24) together with (5.20) into the equation (5.23) and taking the limit $\omega \to \varepsilon_\nu(\boldsymbol{p}) - \mu$, one obtains

$$\left[\varepsilon_\nu(\boldsymbol{p}) + \frac{\hbar^2 \nabla^2}{2m} - v_H(\boldsymbol{r})\right] Z_{\boldsymbol{p}}^{(\nu)}(\boldsymbol{x}, \boldsymbol{x}') -$$

$$-\int d^3 r'' \, \Sigma(\varepsilon_\nu(\boldsymbol{p}) - \mu, \boldsymbol{r}, \boldsymbol{r}'') \, Z_{\boldsymbol{p}}^{(\nu)}(\boldsymbol{r}'', \boldsymbol{r}') = 0. \quad (5.25)$$

Here, Z carries the additional index $\boldsymbol{p}$ since this function is the Fourier transform of the residuum of (5.24) connected with the pole at $\varepsilon_\nu(\boldsymbol{p})$. The variable $\boldsymbol{r}'$ enters the linear equation (5.25) only as a parameter, hence one may put

$$Z_{\boldsymbol{p}}^{(\nu)}(\boldsymbol{r}, \boldsymbol{r}') = \varphi_{\boldsymbol{p}\nu}(\boldsymbol{r}) \, \chi_{\boldsymbol{p}\nu}(\boldsymbol{r}'). \quad (5.26)$$

After substituting this into (5.25) the 'Schrödinger equation'

$$\left[-\frac{\hbar^2 \nabla^2}{2m} + v_H(\boldsymbol{r})\right] \varphi_{\boldsymbol{p}\nu}(\boldsymbol{r}) +$$

$$+ \int d^3 r' \, \Sigma(\varepsilon_\nu(\boldsymbol{p}) - \mu, \boldsymbol{r}, \boldsymbol{r}') \, \varphi_{\boldsymbol{p}\nu}(\boldsymbol{r}') = \varepsilon_\nu(\boldsymbol{p}) \, \varphi_{\boldsymbol{p}\nu}(\boldsymbol{r}) \quad (5.27)$$

is obtained for the quasi-particle spectral amplitude $\varphi_{\boldsymbol{p}\nu}(\boldsymbol{r})$ which replaces the orbital wave function of non-interacting particles [Luttinger, 1961, Bychkov and Gor'kov, 1961]. The dual amplitude $\chi_{\boldsymbol{p}\nu}(\boldsymbol{r})$ obeys the Hermitian conjugate equation.

Neglecting in this equation the term with Σ we would be left with the single-particle Schrödinger equation describing an electron in the mean Hartree field. In reality, the electron is surrounded by an exchange and correlation hole in the distribution of all other electrons. This hole moves together with the considered electron (and it is this combined aggregate that forms the quasi-particle; the exchange and correlation hole in the electron distribution is a depletion of pair density and is not to be confused with hole-type quasi-particles), therefore the exchange and correlation hole depends on the state of motion of the electron, in particular on its energy. (In an inhomogeneous electron system the shape of this hole depends on the electron position $\boldsymbol{r}$, too.) The diameter of this hole is of the order of the density parameter r_s ($4\pi r_s^3/3 = 1/\rho$, where ρ is the electron density). Hence,

$$\Sigma(\varepsilon - \mu, \boldsymbol{r}, \boldsymbol{r}') \approx 0 \text{ for } |\boldsymbol{r} - \boldsymbol{r}'| \gtrsim r_s.$$

Moreover, if the exchange and correlation hole around an electron at position r is expanded in powers of $r' - r$, then the linear terms do not contribute to its energy of interaction with the given electron because of the isotropy of the Coulomb force. (The same holds true for all higher multipole terms.) Finally, we are especially interested in quasi-particle states near the Fermi level. All that suggests for $\varepsilon_\nu(p) \approx \mu$ the local approximation

$$\int d^3r' \, \text{Re} \, \Sigma(\varepsilon_\nu(p) - \mu, r, r') \, \varphi_{p\nu(r')} \approx v_\Sigma(r) \, \varphi_{p\nu}(r) \tag{5.28}$$

for the self-energy operator. Furthermore, neglecting the imaginary part of Σ (which goes to zero for $\varepsilon_\nu(p) \to \mu$), one ends up with the usual Schrödinger equation containing the crystal potential

$$v(r) = v_H(r) + v_\Sigma(r) \tag{5.29}$$

the solution of which should give the band structure (5.14).

Within this frame the 'inverse problem' may be thought of being continued to extract an empirical crystal potential from experimental data. From this angle a great part of the many band structure calculations performed before the eighties of the twenties century must be judged. As mentioned, thereby the quasi-particle damping (imaginary part of Σ) usually is neglected. Keeping in mind the conservation of the total (quasi-) momentum in electron-electron collisions it is easy to show [Landau and Lifshits, 1980b] that at low excitation energies due to the electron-electron interaction

$$\text{Im} \, \varepsilon_\nu(p) \sim (\text{Re} \, \varepsilon_\nu(p) - \mu)^2 \tag{5.30}$$

holds.

In the model of non-interacting with each other particles, the Fermi momentum p_F in the homogeneous liquid (Fig.4a on page 86) or the Fermi surface (5.15) in a crystalline solid respectively are connected with the total particle number N in a very simple way: there must be just N one-particle levels (understanding the multiplicity of possible degeneracies including spin degeneracy) below the Fermi level. If τ_F is the volume in p-space 'inside' of the Fermi surface, that is the volume of the region where $\text{Re} \, \varepsilon_\nu(p) < \varepsilon_F$ and summation over all branches ν of the dispersion law (including the spin index) to be understood, then

$$\frac{N}{V} = \frac{\tau_F}{(2\pi\hbar)^3} \tag{5.31}$$

with the volume V of the system. For real systems of interacting particles, such a connection is not evident by itself. However, Landau's hypothesis, that, for the fermion excitation spectra, (5.31) holds in this general case too, where now N is the number of original particles and τ_F the volume in quasi-momentum space (summed over all branches) corresponding to hole excitations (Fig.4a), was confirmed by [Luttinger, 1960] in a rigorous way and goes now under the name Luttinger theorem.

Note that ^{3}He and metals in the ground state are examples of inhomogeneous (in momentum space) vacua: there is a distinguished surface (the Fermi surface (5.31)), on one side of which particles are created and on the other holes (antiparticles).

5.4 Green's Function of Lattice Phonons

Up to now we have considered the *electronic motion* only in a crystalline lattice with nuclei *fixed* at the lattice sites. The *vibrational motion* of the sites of a harmonic lattice, that is of a lattice in which the potential energy is a quadratic form of the deviations $u_{ns} = s_n - s^0$ of the sites from their equilibrium positions $R_n + s^0$, is the simplest (and minutely studied in textbooks) example of a bosonic quasi-particle spectrum. With the help of the transformation (see for instance [Maradudin, 1974]; we do not explain here in detail the rather standard notation; we continue to treat ω as an energy, not a frequency)

$$\hat{u}_{ns} = \sum_{q\lambda} \left(\frac{e^\lambda_{qs}}{\sqrt{2NM_s}} e^{iqR_n} + c.c. \right) \frac{1}{\sqrt{2\omega_\lambda(q)}} \left(\hat{b}_{q\lambda} + \hat{b}^\dagger_{q\lambda} \right), \qquad (5.32)$$

where the first parentheses contain the transformation to normal coordinates of the vibrational system and the rest of the formula (5.32) expresses the usual transformation of the vibrational system to boson operators, the Hamiltonian of the harmonic lattice is obtained in the form

$$\hat{H}_{\text{ph}} = \sum_{q\lambda} \omega_\lambda(q) \left(\hat{b}^\dagger_{q\lambda} \hat{b}_{q\lambda} + \frac{1}{2} \right). \qquad (5.33)$$

In this formulas, q runs over the first cell of the reciprocal lattice (first Brillouin zone), and λ runs over all polarizations (phonon branches). The quasi-momentum of a phonon is $\hbar q$. (In our earlier notation of Chapter 3, $\alpha \equiv (\lambda, \hbar q)$.) If, in a more general case, the potential energy of the lattice is expanded in a Taylor series in terms of the $\boldsymbol{u}_i$, then, in addition to (5.33), anharmonic terms with products of more than two operators $\hat{b}$ and $\hat{b}^\dagger$ are obtained describing a phonon-phonon interaction. These terms are unimportant at small amplitudes $\boldsymbol{u}_i$, but, as was demonstrated in [Ford, 1975], the presence of analogous terms leads to a chaotic character already of the classical motion of the system at a given energy, so that the stationary quantum states in contrast to the quasi-stationary ones should be expected monstrously complex.

In solid ^{3}He and ^{4}He, already the amplitudes of the zero-vibrations in the ground state are not small in the sense of the harmonic approximation. Having this in mind one speaks of *quantum crystals*. Here, the atoms move in any case in a strongly anharmonic potential. Notwithstanding of that, well established ordinary phonons (5.33) are observed in those crystals. Only their connection with the movement of individual atoms is much more complicated than (5.32) [Guyer, 1969].

The phonon Green's function is usually defined as

$$D_{ns,n's'}(t) = -i\langle|\mathbf{T}\hat{u}_{ns}(t)\hat{u}_{n's'}(0)|\rangle =$$
$$= \frac{1}{2\pi\hbar}\int dt e^{-i\omega t/\hbar} * \qquad (5.34)$$
$$* \frac{1}{\sqrt{M_s M_{s'}}}\frac{V}{(2\pi)^3}\int_{B.Z.} d^3q e^{i\boldsymbol{q}\cdot(\boldsymbol{R}_n - \boldsymbol{R}_{n'})}\sum_\lambda e^{*\lambda}_{\boldsymbol{q}s}D_\lambda(\omega,\boldsymbol{q})e^{\lambda}_{\boldsymbol{q}s'},$$

where $|\rangle$ is now the vacuum of the $\hat{b}$-operators. With (5.32, 5.33) it is an exercise to show that

$$D_\lambda(\omega,\boldsymbol{q}) = \frac{\hbar^2}{\omega^2 - \omega_\lambda(\boldsymbol{q})^2}. \qquad (5.35)$$

In a real solid the vibrational degrees of freedom additionally couple to the degrees of freedom of the electronic motion. On the one hand it leads to a Dyson equation for the phonon Green's function of the form

$$\sum_{n''s''}\left[\sum_\lambda \frac{e^{*\lambda}_{\boldsymbol{q}s}}{\sqrt{M_s}}(\omega^2-\omega_\lambda(\boldsymbol{q})^2)\frac{e^{\lambda}_{\boldsymbol{q}s''}}{\sqrt{M_{s''}}}\delta_{nn''}+\Pi_{ns,n''s''}(\omega)\right]D_{n''s'',n's'}(\omega) = \qquad (5.36)$$
$$= \delta_{nn'}\delta_{ss'},$$

where the self-energy part Π is the polarization operator of the Bloch electrons (diagrammatically the 'bubble' of (A.7), Appendix A I) with two matrix elements (vertices) of the electron-phonon interaction attached to it. On the other hand,

5.4 Green's Function of Lattice Phonons

this electron-phonon coupling leads to a renormalization and damping of low-lying electron modes in a vicinity of the Fermi level of width $\sim \omega_{\max} \sim 30$ meV (see, for instance, [Ashcroft and Mermin, 1976, Chap. 26]).

This electron-phonon interaction is particularly essential in metals where it leads to the most interesting phenomena as to anomalies in the quasi-particle spectra, i.e. in a non-analytical dependence of $\varepsilon_\nu(\boldsymbol{p})$ and $\omega_\lambda(\boldsymbol{q})$ on the quasi-momentum [Migdal, 1958, Kohn, 1959, Kaganov and Lisovskaya, 1981]; see also [Kaganov, 1985] and references therein. It also contributes a phonon part to the electron self-energy Σ of (5.23) in an energy range of the order of the Debye energy (maximal phonon energy) for $|\varepsilon - \mu|$. It leads to a renormalization factor $(1+\lambda) > 1$ for the low temperature electronic specific heat of metals. Just this electron-phonon interaction usually causes the transition of metals into the superconducting state (see for instance [Abrikosov et al., 1975, Landau and Lifshits, 1980b, Chap. V]), implying a qualitative reconstruction of the electronic spectrum.

However, if one ignores these phenomena, e.g. by considering an alkaline metal which does not show superconductivity down to zero temperature, then the electron-phonon interaction does not change the picture qualitatively. Indeed, at low electronic excitation energy one finds

$$\operatorname{Im} \varepsilon_\nu(\boldsymbol{p}) \sim (\operatorname{Re} \varepsilon_\nu(\boldsymbol{p}) - \mu)^3. \tag{5.37}$$

due to electron-phonon interaction [Migdal, 1958].[32] In these estimates (5.30) and (5.37), first of all it is assumed that besides the considered one there are no further quasi-particles present, that is, the system should be at $T = 0$. Second, a decisive role in these estimates is played by the conservation of quasi-momentum in the scattering and reaction processes of quasi-particles limiting the possibility of those processes severely. Thus, in an ideal metallic crystal at $T = 0$ in the normal (non-superconducting) state there are *stable quasi-particles* on the Fermi surface, the corresponding Green's function in fact having a pole on the real axis of the complex energy plane:

$$\operatorname{Im} \varepsilon_\nu(\boldsymbol{p}_F) = 0. \tag{5.38}$$

For the same reason the electron at the bottom of the conduction band and the hole at the top of the valence band, respectively, of an ideal crystalline semiconductor at

---

[32] There appear papers in the literature from time to time asserting that a metal would become unstable for an electron-phonon coupling strength $\lambda > 1$. These results are based on an incorrect field-theoretic treatment of Frölich's model Hamiltonian and are not valid [Maksimov, 1976, Brovman and Kagan, 1974]. This incorrectness is also contained in Migdal's original paper without, however, corrupting the results we refer to.

$T = 0$ are stable quasi-particles. Generally speaking, in a (discrete) translationally invariant system the Fermi particle and hole states of lowest excitation energy are always stable. They correspond to the lowest energy *stationary* state of a system of $N \pm 1$ original particles with a total (quasi-)momentum $\boldsymbol{p}$.

On the other hand, at $T \neq 0$ or when defects are present, $\operatorname{Im} \varepsilon_\nu(\boldsymbol{p})$ preserves a finite non-zero value everywhere. (Im ε_α depends on the number of quasi-particles present and thus on temperature as it was the case for Re ε_α in (5.3).)

5.5 Disordered Systems

More complicated than described above is the situation in disordered systems. Here, the momentum or quasi-momentum of the quasi-particle cannot be introduced from the outset, though it may prove to be a useful approximate quantum number for electronic excitations of sufficiently high energy (so that their kinetic energy dominates) or for long-wavelength elastic waves. Even the quasi-particle vacuum cannot be introduced because we do not have a set of Fock space operators to define it.

The equation (5.23) for the one-particle Green's function makes no use of the translational invariance and hence may be carried over to the considered case. We cast it into the operator form

$$[\omega + \mu - \hat{H}(\omega)] \hat{G}(\omega) = \hbar \hat{I}, \tag{5.39}$$

where $\hat{I}$ is the unit operator,

$$\hat{H}(\omega) = -\frac{\hbar^2 \nabla^2}{2m} + v_H + \hat{\Sigma}(\omega) \tag{5.40}$$

and $\hat{\Sigma}$ and $\hat{G}$ are integral operators with kernels $\Sigma(\omega, \boldsymbol{r}, \boldsymbol{r}')$ and $G(\omega, \boldsymbol{r}, \boldsymbol{r}')$, respectively. For small ω the expansion

$$\hat{\Sigma}(\omega) \approx \hat{\Sigma}_0 + \omega \hat{\Sigma}', \quad \Sigma'(\boldsymbol{r}, \boldsymbol{r}') \approx \sigma \, \delta(\boldsymbol{r} - \boldsymbol{r}') \tag{5.41}$$

may be considered, assuming that in the linear term a local approximation makes sense. From (5.39) we now have

$$\hat{G}_0(\omega) = \hbar \left[(1 - \sigma)\omega + \mu - \hat{H}_0 \right]^{-1}, \quad \hat{H}_0 = -\frac{\hbar^2 \nabla^2}{2m} + v_H + \hat{\Sigma}_0 \tag{5.42}$$

containing the frequency renormalization factor $(1 - \sigma)$. From this expression, the simple form (5.24) or (3.32) of the pole term cannot be extracted, because

5.5 Disordered Systems

there is no simple representation diagonalizing the expression (5.42) (as the α-representation of (3.32)) or algebraizing it (as the p-representation of (5.24)). However, in the spirit of those terms, for the quasi-particle density of states[33] normalized to the volume V of the system one always may write down

$$\begin{aligned} D_{\text{qp}V}(\varepsilon - \mu) &= -\frac{1}{\pi V} \operatorname{Im} \operatorname{tr}[\hat{Z}^{-1}\hat{G}_0](\varepsilon - \mu + i0) = \\ &= -\frac{1}{\pi V} \operatorname{Im} \int_V d^3 r \, [Z^{-1}G_0](\varepsilon - \mu + i0, \boldsymbol{r}, \boldsymbol{r}). \end{aligned} \quad (5.43)$$

In the disordered case it is not clear from the outset how this quantity fluctuates in dependence of the choice of the cut-out volume V of the macroscopic system or equivalently, in a thermodynamic description, in dependence of the member of the statistical ensemble describing the disorder. A physically observable quantity should be *self-averaging*, that is,

$$D_{\text{qp}}(\varepsilon - \mu) = \lim_{V \to \infty} D_{\text{qp}V}(\varepsilon - \mu) \quad (5.44)$$

should be unambiguously defined. (In other words, the limes (5.44) should exist.) For the quasi-particle density of states the self-averaging property has been proved under very general assumptions [Lifshits et al., 1988]. The equations (5.43) and (5.42) then lead to the problem of finding the spectrum of a stochastic one-particle Hamiltonian $\hat{H}_0$ the 'eigenstates' of which may have the character either of running waves or of localized states.

The alternative quasi-particle renormalization parameters Z and $(1 - \sigma)$ have been used to connect the quasi-particle theory with the scaling theory of the metal-insulator transition [Castellani et al., 1987]. Without going into details we only mention that, starting from the metallic side ($D_{\text{qp}}(0) \neq 0$), we may approach the metal-insulator transition ($D_{\text{qp}}(0) \to 0$), if $(1 - \sigma) \to \infty$. Starting from the insulating side, an alternative approach to the transition would be $Z \to 0$.

An important special case is that of a 'Coulomb glass', localized charged states at spatially fixed centers of a disordered semiconductor, where the potential values U_i at centers $\boldsymbol{x}_i$ fluctuate from position to position. The total energy of the system

---

[33]That is the number of quasi-particle states per energy interval. Also in the present context this quantity has of course nothing in common with (2.35). In particular, $\operatorname{Im} \operatorname{tr}[\hat{Z}\hat{G}]$ increases linearly with the size of the system (with the volume V), whereas $D(E)$ of (2.35) increases exponentially.

as a function of the occupation numbers $n_i = 0, 1$ is in the simplest case

$$E(n_i) = \sum_i U_i n_i + \frac{1}{2} \sum_{i \neq j} \frac{n_i n_j}{\epsilon |\boldsymbol{x}_i - \boldsymbol{x}_j|}, \tag{5.45}$$

where only in this formula (5.45) and in (5.46) below, ϵ denotes the dielectric constant of the semiconductor. According to (5.2), the quasi-particle energy is

$$\operatorname{Re} \varepsilon_i = U_i + \sum_{j(\neq i)} \frac{n_j}{\epsilon |\boldsymbol{x}_i - \boldsymbol{x}_j|}. \tag{5.46}$$

Again we are interested in the quasi-particle density of states $D(\varepsilon)$. Although the relations (5.45 and 5.46) look very simple, the problem of finding the corresponding spectrum $D(\varepsilon)$ is very complicated [Shklovskii and Efros, 1984, Efros and Shklovskii, 1985]. Its solution starts with the determination of the ground state of N_e 'original' electrons to be distributed over sites $\boldsymbol{x}_i$ with given potential energies U_i. (Of course, the number of sites must exceed the number of electrons.) In other words, one must find the numbers n_i for which (5.45) takes its minimum value. This set must fulfill (among a lot of others) the condition that all levels $\varepsilon_i^{(0)}$ of (5.46) corresponding to sites where $n_i = 0$ must be above all levels $\varepsilon_i^{(1)}$ at sites with $n_i = 1$, even stronger: $\varepsilon_i^{(0)} - \varepsilon_i^{(1)} - 1/\epsilon|\boldsymbol{x}_i - \boldsymbol{x}_j| > 0$, otherwise energy could be gained by a redistribution. Above all $\varepsilon_i^{(1)}$ and below all $\varepsilon_i^{(0)}$ the Fermi level μ is placed. The in this way obtained sets $\{n_i\}$ and $\{\varepsilon_i\}$ correspond to given sets $\{\boldsymbol{x}_i, U_i\}$. These results must then be averaged over a probability distribution of sets $\{\boldsymbol{x}_i, U_i\}$. Usually the influence of the distribution of sites $\boldsymbol{x}_i$ is assumed to be less important, and for the sake of simplicity one uses a regular lattice of vectors $\boldsymbol{x}_i$.

No step of this solving procedure can be performed analytically, although it seems possible to estimate from above the density of states $D(\varepsilon)$ near the Fermi level in an analytical way. Already these estimates show that, due to the long-range Coulomb forces, a 'soft' gap opens up in the spectrum at the Fermi level, called a Coulomb gap, and that the actual form of the quasi-particle density of states there depends significantly on the dimension d of the system (dimension of the lattice of vectors $\boldsymbol{x}_i$). For $d > 1$,

$$D(\varepsilon) \approx |\varepsilon - \mu|^\nu \tag{5.47}$$

seems to hold at least for not too small $|\varepsilon - \mu|$, but reliable values of ν are only obtained from rather sophisticated and time-consuming numerical computations

[Baranovskii et al., 1979, Davies et al., 1984] giving $\nu > 2$ for $d = 3$ and $\nu \approx 1.5$ for $d = 2$. For $d = 1$, a careful computer analysis [Möbius and Richter, 1986] gave

$$D(\varepsilon) \sim -\ln^{-1}|\varepsilon - \mu|. \tag{5.48}$$

The existence of the Coulomb gap plays an important role in the description of electrical transport properties of heavily doped semiconductors. An interesting point is that the Coulomb gap is not seen in the caloric properties (e.g. specific heat). The simple explanation is that the one-particle excitations (5.46) do not exhaust the excitation spectrum of the solid: excited states within the Coulomb gap may be obtained by *simultaneously* redistributing two or more neighboring electrons [Efros and Shklovskii, 1985]. It can be shown, however, that these excitations do not contribute to the electrical dc conductivity which hence is governed by the Coulomb gap.

5.6 Frontiers of the Quasi-Particle Picture

Finally we give two examples where the quasi-particle picture runs obviously into difficulties: Let some ions move in a double-well potential [Anderson et al., 1972], where the height of the barrier between the two wells does not allow tunneling within any measuring time. Then, at sufficiently low temperatures, each ion is captured in one of the wells, but *the levels are not occupied in accordance with statistics*. As the temperature is raised, thermally activated hopping becomes possible and the occupation of the levels thermalizes. The thermalization cannot, however, be described by elementary collision acts. The disorder leads to a continuous (at least in the scale of temperature) energy distribution of the levels,[34] displaying experimentally in a large linear (in temperature) term in the heat capacity [Anderson et al., 1972] and in a complicated behavior of the ac conductivity [Pollak and Pike, 1972]. Due to the fact that the *kinetics of thermalization does not go via elementary collision acts*, the excitations in question do not take normally part in the heat transport, but only scatter phonons thus increasing the thermal resistivity [Gurevich and Parashkin, 1982].

Another even more illuminating example is given by the role of f-levels of rare-earth ions in metals (e.g. [Khomskii, 1979, Fulde, 1995]). Let $E(n_f, n_p)$ be the total energy of the system, depending (in the spirit of Landau's Fermi-liquid theory) on the occupation numbers of localized f-states and of band states with

---
[34] That is, many levels on an interval $\Delta\varepsilon \sim T$.

quasi-momentum $\boldsymbol{p}$. Let the total number of electrons be fixed, so that a change of n_f implies an opposite change of $n_{\boldsymbol{p}_F}$. If we fix the energy scale so that $\varepsilon_F = \mu = 0$, we may consider E to be a function of n_f alone ($E = E(n_f)$). In analogy with (5.2), let us *formally* write

$$\varepsilon_f(n_f) = \frac{dE(n_f)}{dn_f}. \tag{5.49}$$

Because of the strong localization of the f-state, a change of n_f strongly influences the electron density, manifesting in a strong non-linearity of the function $E(n_f)$. It may happen that $\varepsilon_f(0) < 0$, but $\varepsilon_f(1) > 0$. If the f-state does not hybridize with the band states, then the state Ψ of the system should be a linear combination of states Φ_0 and Φ_1 being eigenstates of the occupation number operator (note that Ψ and Φ are many-body states of the whole electron system):

$$\hat{n}_f \, \Phi_0 = 0, \quad \hat{n}_f \, \Phi_1 = \Phi_1, \quad \langle \Phi_0 | \Phi_1 \rangle = 0. \tag{5.50}$$

Quite generally, let

$$\Psi(n_f) = e^{i\alpha} \sqrt{1 - n_f} \, \Phi_0 + \sqrt{n_f} \, \Phi_1 \tag{5.51}$$

with

$$\langle \Psi | \Psi \rangle = 1; \quad \langle \Psi | \hat{n}_f | \Psi \rangle = n_f, \tag{5.52}$$

but

$$\langle \Psi | \hat{n}_f^2 | \Psi \rangle = n_f^2 < n_f \quad \text{for } 0 < n_f < 1. \tag{5.53}$$

Clearly, the given situation describes a *valence fluctuation* $\langle \hat{n}_f^2 \rangle - \langle \hat{n}_f \rangle^2 > 0$. Furthermore,

$$\begin{aligned} E(n_f) &= \langle \Psi | \hat{H} | \Psi \rangle = \\ &= (1 - n_f) \, E(0) + n_f \, E(1) + 2\sqrt{n_f(1 - n_f)} \, H, \end{aligned} \tag{5.54}$$

where the phase α must be chosen so that

$$H \equiv \operatorname{Re}\left[e^{-i\alpha} \langle \Phi_0 | \hat{H} | \Phi_1 \rangle\right] = -|\langle \Phi_0 | \hat{H} | \Phi_1 \rangle| \tag{5.55}$$

in order to yield the minimum value of (5.54) for any given n_f.

5.6 Frontiers of the Quasi-Particle Picture

The ground state of the system is determined by the equations

$$\left.\frac{dE(n_f)}{dn_f}\right|_{n_f^{(0)}} = 0, \quad E_0 = E(n_f^{(0)}). \tag{5.56}$$

These relations do, however, not describe a quasi-particle with $\varepsilon_f(n_f^{(0)}) = 0$, since there is no corresponding excited state: The states $\Psi(n_f)$ and $\Psi(n_f')$ are orthogonal to each other only, if $n_f = 0$ and $n_f' = 1$. Only these numbers (0 or 1) may be found in experiment, if the valence of a chosen single ion in the metal is directly measured (e.g. via an NMR level shift, if the occupation fluctuation is slow enough compared to the measuring time).

If, however, the f-states hybridize forming an albeit very narrow band (as in the heavy fermion case; [Steward, 1984, Brandt and Moshkalov, 1984, Fulde, 1995, Hewson, 1993], then, contrary to (5.51), the state $\Psi(n_f)$ is now obtained by partially filling the band, and consequently all $\Psi(n_f)$ are orthogonal to each other. In this case, $\varepsilon_f = dE/dn_f$ defines a quasi-particle, and, moreover, the total-energy minimum corresponds either to $n_f = 0$, or to $\varepsilon_f = 0$, or to $n_f = 2$ (note that in these structures there is always a center of inversion so that each electronic level is twofold Kramers-degenerate). These considerations do of course not answer the most important question: why is the ground state in one case of the form (5.51) and in the other case of the band character despite of the very small bandwidth.

The difficulties with the quasi-particle conception in the above considered examples originate mainly from the fact that in these cases the interaction energy is larger than the quasi-particle energy itself.

6 The Nature of the Vacuum

The energy of a given quasi-particle depends, due to interaction, on the other quasi-particles present and hence on the state of the system. This influence of the state of the system on the quasi-particle spectrum is for instance expressed by the equation (5.3). For a thermodynamic equilibrium state it may be obtained from Green's function theory at non-zero temperature (Section 3.7) in complete analogy to the $T = 0$ case.

In many cases this describes not the only and not the most important influence of the state of the system. In Chapter 3 we assumed for the sake of simplicity that the quasi-particle vacuum Ψ_0 would be the ground state of the system. Also if one realizes that in certain cases at $T = 0$ this may indeed be true (but remember the remarks at the beginning of Chapter 3), with rising temperature the system may undergo a phase transition, thereby its state and the quasi-particle spectrum being reconstructed. This means already not a gradual change of the renormalization and damping of ε_α due to the presence of the term $\hat{H}_{\text{int}}$ in (4.65) which was expressed by (5.3) but a drastic change of the spectrum which of course is also caused by $\hat{H}_{\text{int}}$, which, however, manifests itself usually in a change of (internal or external) symmetry of the state of the system including a change of the symmetry of Ψ_0.

Note also that it is the polarizability of the vacuum that prevents a many-body wave function of quasi-particles from making sense: if a quasi-particle of sufficient energy is present in a solid to shake up an electron-hole pair or a lattice phonon, then it creates these new particles by loosing energy. The quasi-particle number and hence the number of quasi-particle degrees of freedom on which a quasi-particle many-body wave function would depend is not conserved, and a Fock space description is mandatory. The same reason prevents a Schrödinger-like wave function from existing in relativistic field theory.

6.1 Spontaneous Symmetry Breaking

Consider as an example a superconductor. Below the transition temperature T_c the vacuum state Ψ_0 contains a condensate while above T_c it does not. The Hamiltonian of the system is gauge invariant while the ground state is not (see next section for details). In this case the gauge symmetry for the fermion field describing the

6.1 Spontaneous Symmetry Breaking

electronic quasi-particles is spontaneously broken below T_c in the ground state resulting in a particular behavior of the ground state in an applied magnetic field and in most cases in a gap in the electron-hole spectrum. In a structural transition the external crystal symmetry of the state Ψ_0 changes thereby the quasi-particle spectrum getting another periodicity. In an amorphous solid, Ψ_0 is at least in most cases supposed not to be the ground state even at $T = 0$, the latter being assumed crystalline. (Both statements cannot at present be proved 'from the first principles.') It should also be mentioned that the spontaneous breaking of a symmetry in Ψ_0 (that is, the symmetry of Ψ_0 is lower than that of the initial Hamiltonian of the system) causes the presence of a special bosonic quasi-particle branch, an acoustic (Goldstone) or optic (Higgs) mode in dependence on the actual situation (see, e.g., [Grib et al., 1971]). These considered circumstances destroy the naive picture according to which the quasi-particle vacuum represents an empty vessel which the quasi-particles are poured into and through which they move, possibly interacting with each other, and in this way mutually influencing their dynamical characteristics (energy and lifetime). In truth the vacuum is a physical reality with a complex inner structure which determines the motion of particles but on which the particles present act back determining its structure (bootstrap principle).[35]

In order to understand this in more detail we continue a technical analysis from Section 2.1.

The point is that the interpretation of the field operators $\hat{a}_\alpha^\dagger$ and $\hat{a}_\alpha$, the $\hat{c}$-operators of (4.2) or the $\hat{b}$-operators of (4.15), as creation and annihilation operators is based on the Fock space representation of these quantities. The vacuum state is the cyclic state of this representation:

$$\hat{a}_\alpha|\Psi_0\rangle = 0 \text{ for all } \hat{a}_\alpha. \tag{6.1}$$

and the representation space is the Fock space

$$\mathcal{F} = \{ \ |\Psi\rangle = R(\hat{a}_\alpha^\dagger)|\Psi_0\rangle \ | \ \langle\Psi|\Psi\rangle < \infty \ \} \tag{6.2}$$

the geometry of which as a Hilbert space is determined in accordance with (6.1), with the mutual conjugation of $\hat{a}_\alpha^\dagger$ and $\hat{a}_\alpha$, and with the canonical (anti-)commutation relations of these operators. In (6.2), $R(\hat{a}_\alpha^\dagger)$ means any power series of operators $\hat{a}_\alpha^\dagger$, for which $\langle\Psi|\Psi\rangle$ is finite. Since different monomials of the series $R(\hat{a}_\alpha^\dagger)|\Psi_0\rangle$ are orthogonal to each other, the finiteness of the Hermitian norm

---

[35] A 'principle' introduced into particle physics, according to which somebody is lifting up himself by pulling his own bootstraps.

of $|\Psi\rangle$ means that this vector may be approximated with any accuracy by *polynomial* expressions $P(\hat{a}_\alpha^\dagger)|\Psi_0\rangle$. Accordingly, in a Fock space an arbitrarily large but always finite number (as the expectation value of the number operator) of particles may only be created. This restriction is necessary in order that the quantities $\hat{a}_\alpha^\dagger$ and $\hat{a}_\alpha$ and hence all microscopic observables are correctly defined as operators in the Fock space. (They are defined on the dense subset of $\mathcal{F}$ consisting of all polynomial states $P(\hat{a}_\alpha^\dagger)|\Psi_0\rangle$.)

Suppose now that in a macroscopic system a state $|\Psi\rangle$ being homogeneous in space changes by some process into another state $|\Psi'\rangle$ being again homogeneous in space. If one divides the volume V of the system into (still macroscopic) partial volumes V_i, $i = 1, \ldots, N$ and if one describes the change of the state in each partial volume V_i by $\Psi'_{V_i} = R_{V_i}(\hat{a}_\alpha^\dagger, \hat{a}_\alpha)|\Psi_{V_i}\rangle$, where $R_{V_i}(\hat{a}_\alpha^\dagger, \hat{a}_\alpha)$ is a power series of operators $\hat{a}_\alpha^\dagger$ and $\hat{a}_\alpha$ localized in V_i, then (neglecting surface corrections at the surfaces of the macroscopic volumes V_i) one obtains for the whole system

$$|\Psi'\rangle = \prod_{i=1}^{N} R_{V_i}(\hat{a}_\alpha^\dagger, \hat{a}_\alpha)\,|\Psi\rangle. \tag{6.3}$$

Consider here the thermodynamic limit $N \to \infty$. If $|\Psi\rangle$ was a state in the Fock space, then according to (6.2) $|\Psi'\rangle$ is usually already no longer in that Fock space (Grib *et al.* 1970). Since for normalized states $\langle \Psi'_{V_i}|\Psi_{V_i}\rangle < 1$, it follows

$$\langle \Psi'|\Psi\rangle = \lim_{N\to\infty} \langle \Psi'_{V_i}|\Psi_{V_i}\rangle^N = 0, \tag{6.4}$$

that is, different macroscopically homogeneous states are always orthogonal to each other. Compare also the considerations of Section 2.1.

Let

$$\hat{A}(\boldsymbol{x}) = R_A(\hat{a}_\alpha^\dagger, \hat{a}_\alpha) \tag{6.5}$$

be any local observable localized in the a small vicinity of the point $\boldsymbol{x}$ and let $\hat{B}(\boldsymbol{x})$ be a (bosonic) quantity of the same local type having different values in the states $|\Psi\rangle$ and $|\Psi'\rangle$, Then

$$\hat{b} = \lim_{V\to\infty} \frac{1}{V} \int_V d^3x\, \hat{B}(\boldsymbol{x}) \tag{6.6}$$

is a global bosonic density operator like (2.4). Since (no matter whether $\hat{A}$ is bosonic or fermionic) the commutator $[\hat{B}(\boldsymbol{x}'), \hat{A}(\boldsymbol{x})]_- = 0$ at sufficiently large distances

$|\boldsymbol{x} - \boldsymbol{x}'|$, the integral in the following expression remains finite while extending the volume, hence [Haag, 1962]

$$[\hat{b}, \hat{A}(\boldsymbol{x})]_- = \lim_{V \to \infty} \frac{1}{V} \int_V d^3x \, [\hat{B}(\boldsymbol{x}'), \hat{A}(\boldsymbol{x})]_- = 0. \tag{6.7}$$

From the preceding considerations it follows further that a change of the value of $b = \langle \hat{b} \rangle$ is necessarily connected with abandoning the Fock space.

Consequently, as already discussed in Section 2.1, in the thermodynamic limit the observables split into two distinct classes: local (microscopic) observables, as previously expressed by field operators in the Fock space, and global (macroscopic) observables as e.g. the mean particle density, the averaged magnetization, the condensate density, and so on, the formal constructions of which by field operators commute with all local operators and have expectation values constant in the whole Fock space in accordance with the construction of the latter. A change of the state of the macrosystem including a change of extensive observables corresponds to a transition from the Fock-space representation of the field operators to another inequivalent (non-Fock-space) representation [Grib et al., 1971, Emch, 1972]. According to the general mathematical representation theory of normed algebras (the Gelfand-Naimark-Segal construction theorem, see Section 2.2), with the new state again a cyclic representation of field operators in a Hilbert space may be associated, i.e. such a representation that all vectors of the Hilbert space may be approximated by acting with field operators on only one (the cyclic) vector of the space [Haag and Kastler, 1964].

6.2 Gauge Symmetry

Let us consider in some more detail the case of charged particles interacting with a gauge field. The simplest case is a system of non-relativistic particles of charge q interacting with an electromagnetic field, $\boldsymbol{B} = \nabla \times \boldsymbol{A}$, $\boldsymbol{E} = -\partial \boldsymbol{A}/\partial t - \nabla U$, described by a vector potential $\boldsymbol{A}$ and a scalar potential U. The general case is a Yang-Mills field with particle charges attached to the generators of the compact semi-simple Lie group associated with the Yang-Mills field [Itzykson and Zuber, 1980]. The interaction is obtained from a general principle which in analogy to Einstein's principle of general relativity was termed 'the relativity principle in the charge space' by Hermann Weyl in 1929. For our simple case it means that the charge q is a scalar associated to the one-dimensional representation of the group $U(1)$ of rotations in the complex plane. The gradient $\nabla \hat{\psi}(\boldsymbol{r}, t)$ of the complex field $\hat{\psi}(\boldsymbol{r}, t)$ is to be replaced by the 'covalent gradient' $\boldsymbol{D}\hat{\psi}(\boldsymbol{r}, t) = (\nabla - iq\boldsymbol{A}/\hbar)\hat{\psi}(\boldsymbol{r}, t)$ while $\partial \hat{\psi}(\boldsymbol{r}, t)/\partial t$

is to be replaced by $D_t\hat{\psi}(\boldsymbol{r},t) = (\partial/\partial t + iqU/\hbar)\hat{\psi}(\boldsymbol{r},t)$. This construct is invariant against Lie group transformations of the charge space, in our case

$$\boldsymbol{A}(\boldsymbol{r},t) \longrightarrow \boldsymbol{A}(\boldsymbol{r},t) + \nabla\chi(\boldsymbol{r},t), \quad U(\boldsymbol{r},t) \longrightarrow U(\boldsymbol{r},t) - \frac{\partial\chi(\boldsymbol{r},t)}{\partial t}, \quad (6.8)$$
$$\hat{\psi}(\boldsymbol{r},t) \longrightarrow \hat{\psi}(\boldsymbol{r},t)e^{iq\chi(\boldsymbol{r},t)/\hbar}$$

with an arbitrary gauge function χ (local gauge in space and time). It connects the scalar gauge field and the longitudinal part of the vector gauge field with 'rotations' of the particle field in the charge space. For instance in the 'wave gauge' the scalar potential can be completely gauged away: $U = 0$, $\boldsymbol{E} = -\partial\boldsymbol{A}/\partial t$. This is particularly convenient in stationary situations (see below).

From the second part of (6.8) it is immediately seen that any local sesquilinear form of the particle field $\hat{\psi}$, that is, any expression $\hat{\psi}^\dagger(\cdots)\hat{\psi}$ with no differentiation in the parentheses is invariant under rotations in the charge space. (For a general Yang-Mills field, $\hat{\psi}^\dagger$ is to be replaced by $\hat{\bar{\psi}}$ which is both Hermitian conjugate to $\hat{\psi}$ and forming the charge group representation adjoint to that formed by $\hat{\psi}$.) It is natural to postulate that the vacuum state $|\rangle$ with no charge present is invariant under rotations in the charge space. Recall that any one-particle state is created out of the vacuum as $|1\rangle = \hat{a}_i^\dagger|\rangle$, with $\hat{a}_i^\dagger = \int dx\, \phi_i(x)\hat{\psi}^\dagger(x)$, following from (1.140) and (1.144). Then, since $\hat{a}_i^\dagger$ does not contain a phase (cf. (1.86) and (1.92)), a gauge transformation implies $\phi_i(x) \longrightarrow \phi_i(x)e^{iq\chi(\boldsymbol{r},t)/\hbar}$. In particular, a *global gauge transformation* with a homogeneous $\chi =$ const. rotates any state $|\Psi\rangle$ with total charge Q according to

$$|\Psi\rangle \longrightarrow |\Psi\rangle e^{-iQ\chi/\hbar}, \quad Q = \langle\Psi\left|\int d^3r\, \hat{\psi}^\dagger(\boldsymbol{r})q\hat{\psi}(\boldsymbol{r})\right|\Psi\rangle \quad (6.9)$$

(A possible spin structure is dropped in the second relation.)

These considerations clearly demonstrate, that *any state with non-zero total charge is not invariant with respect to rotations in the charge space.*

Consider now the transformation of Section 4.1 from primary charged particles to quasi-particles. It splits the Hamiltonian (4.7), the operator of total charge (4.8) and any observable into a macroscopic c-number part and a quasi-particle operator. Besides, the charge is 'renormalized' to be q for quasi-particles and $-q$ for quasi-holes. The ground state, originally filled with a non-zero density of primary particles, is the vacuum of the quasi-particle field (4.9) upon which the quasi-particle Fock space is built. It is now possible to consider separately the interaction of the classical c-number part of the system, which does not participate in the quantum dynamics, and of the quasi-particle part with the gauge field. This introduces rotations in a quasi-particle charge space. Now, *the quasi-particle vacuum is again invariant under global gauge transformations.* The breaking of the gauge symmetry in the ground state of primary particles is 'renormalized' away in the quasi-particle description. Observe also that the *Green's function (5.22) is never invariant under local gauge transformation (6.8), except for $r' = r$* which yields the density. For a time independent gauge function (for instance in situations with time independent U), it transforms according to

$$G(t, r, r') \longrightarrow G(t, r, r')e^{iq(\chi(r)-\chi(r'))/\hbar}, \tag{6.10}$$

if the quasi-particle vacuum $|\rangle$ is invariant.

Next, consider the transformations (4.36, 4.37) and recall that here ζ and ξ distinguish the spin, not the charge which is equal for both fields (positive or negative depending on $|k|$). This is connected with the fact that there is a condensate in the ground state of a superconductor which serves as a reservoir out of which pairs of electrons or pairs of holes may be created, while in the normal state only electron hole pairs (of zero total charge) may be created. Hence, in the superconducting state the total quasi-particle charge is not a conserved quantum number any more (while of course the total primary electron charge is and the system must be described 'Weyl-covariant' in the primary quantities: primary observables must be invariant under local gauge transformations. A further consequence is that in the superconducting state the electron Green's function splits into a part which transforms as (6.10) and another (anomalous) part F which transforms according to

$$F(t, r, r') \longrightarrow F(t, r, r')e^{iq(\chi(r)+\chi(r'))/\hbar} \tag{6.11}$$

as well as its transposed and Hermitian conjugated, the diagonal part (that is $r = r'$) of which yields the condensate density, the order parameter of the superconducting state. These circumstances characterize the kind of spontaneous gauge symmetry breaking in the transition into superconductivity.

6.3 Anomalous Mean Values

In the Fock space the cyclic vector is the vacuum vector, for which $\langle|\hat{a}_\alpha^\dagger|\rangle = 0 = \langle|\hat{a}_\alpha|\rangle$. Such relations also hold true in any occupation number eigenstate (1.87) or (1.93), but in a general state this need not be true. For instance in a coherent state (1.98) of a bosonic system there are *anomalous mean values*

$$\langle b|\hat{b}_\alpha^\dagger|b\rangle \neq 0 \neq \langle b|\hat{b}_\alpha|b\rangle. \tag{6.12}$$

Recall that in the bosonic case $\hat{b}_\alpha + \hat{b}_\alpha^\dagger$ may describe an observable.

In the fermionic case, only products of an even number of field operators may describe observables. For the Hamiltonian (4.1), if there are states with $\varepsilon_\alpha - \mu < 0$, in the ground state $|\Psi_0\rangle$ the average charge density is non-zero:

$$\lim_{V\to\infty} \frac{1}{V} \sum_\alpha \langle \Psi_0|\hat{c}_\alpha^\dagger q \hat{c}_\alpha|\Psi_0\rangle \neq 0, \tag{6.13}$$

while the average quasi-particle charge density $\lim_{V\to\infty} V^{-1}\langle \hat{Q}_\mu^p\rangle$ of (4.8) is zero in the whole Fock space created by the quasi-particle operators $\hat{\zeta}_\alpha^\dagger$ and $\hat{\xi}_\alpha^\dagger$ out of the vacuum $|\rangle$ of (4.6):

$$\lim_{V\to\infty} \frac{1}{V}\langle \Psi|\hat{Q}_\mu^p|\Psi\rangle = 0 \text{ for all } |\Psi\rangle. \tag{6.14}$$

(The Fock-space representation of the operators $\hat{c}_\alpha^\dagger$ and $\hat{c}_\alpha$ would correspond to a zero density of primary particles.) In the thermodynamic limit considered in Section 2.1, the state $|\Psi_0\rangle$ of (6.13) leaves the Fock space, and in view of (6.14), (6.13) is also considered to be an anomalous mean value.

An increase of the particle density (and hence of μ to a higher value μ') in (4.4) implies a transformation

$$\hat{\zeta}_\alpha' = s_\alpha \hat{\xi}_\alpha^\dagger \text{ for } \mu < \varepsilon_\alpha < \mu'. \tag{6.15}$$

This is a non-linear transformation (the Hermitian conjugation is a non-linear operation), and it is connected with a change of the quasi-particle spectrum $|\varepsilon_\alpha - \mu|$.

A non-trivial example is given by the Bogolubov-Valatin transformation (4.36, 4.37) in the BCS theory of superconductivity which is connected with anomalous mean values of the types $\langle \hat{\psi}^\dagger \hat{\psi}^\dagger \rangle$ and $\langle \hat{\psi}\hat{\psi}\rangle$ and in most cases (at least nearly everywhere on the Fermi surface) with the occurrence of a gap in the quasi-particle spectrum.

The thermal expansion of a crystal in the quasi-harmonic approximation yields a non-trivial example for a boson field. If (5.32) at $T = 0$ describes the quantized

6.3 Anomalous Mean Values

vibrations $\boldsymbol{u}_i = \boldsymbol{R}_i - \boldsymbol{R}_i^0$ of the lattice points around their mean positions $\boldsymbol{R}_i^0$, then, at $T > 0$ different vibrations

$$\hat{u}_{ns}^T = \sum_{q\lambda} \left(\frac{e_{qs}^{T\lambda}}{\sqrt{2NM_s}} e^{i\boldsymbol{q}\boldsymbol{R}_n^T} + \text{c.c.} \right) \frac{1}{\sqrt{2\omega_\lambda^T(\boldsymbol{q})}} \left(\hat{b}_{q\lambda}^T + \hat{b}_{q\lambda}^{T\dagger} \right) \tag{6.16}$$

are found around different centers $\boldsymbol{R}_n^T$ with different polarization vectors $e_{qs}^{T\lambda}$ and with different frequencies $\omega_\lambda^T(\boldsymbol{q})$ (also the range of $\boldsymbol{q}$-values changes). Because $\boldsymbol{u}_{ns}^T = \boldsymbol{s}_n - \boldsymbol{s}^{0T}$, one has

$$\boldsymbol{u}_{ns}^T = \boldsymbol{u}_{ns} + (\boldsymbol{s}^0 - \boldsymbol{s}^{0T}). \tag{6.17}$$

Moreover, a linear correspondence between the $\boldsymbol{q}$-values of (5.32) and those of (6.16) may be established (associated with the linear correspondence between the $\boldsymbol{R}_n$ and the $\boldsymbol{R}_n^T$). For clarity in the next equation the $\boldsymbol{q}$-values corresponding to (5.32) and (6.16), respectively, are denoted as $\boldsymbol{q}$ and $\tilde{\boldsymbol{q}}$. Then, (6.17) is equivalent to a certain transformation of the form[36]

$$\hat{b}_{\tilde{q}\lambda}^T = \sum_{q'\lambda'} \left(U_{qq'}^{\lambda\lambda'}(T)\, \hat{b}_{q'\lambda'} + V_{qq'}^{\lambda\lambda'}(T)\, \hat{b}_{q'\lambda'}^\dagger \right) + \beta_\lambda(\boldsymbol{q}, T) \tag{6.18}$$

with c-numbers β_λ. The changes of $\boldsymbol{R}_n^T$, $\omega_\lambda^T(\boldsymbol{q})$, and $e_{qs}^{T\lambda}$ are caused by a change of the mean occupation numbers of interacting phonons in the system as temperature rises.

---

[36] The transformations (4.4, 6.15), (4.36, 4.37) and (6.18) are special cases of a class of *canonical transformations* [Berezin, 1965]

$$\hat{a}'_\alpha = \int d\beta \, (U_{\alpha\beta} \hat{a}_\beta + V_{\alpha\beta} \hat{a}_\beta^\dagger) + f_\alpha$$

of the canonical (anti-)commutation relations, where for boson and fermion operators

$$\int d\gamma \, (U_{\alpha\gamma} U_{\beta\gamma}^* \mp V_{\alpha\gamma} V_{\beta\gamma}^*) = \delta_{\alpha\beta}, \quad \int d\gamma \, (U_{\alpha\gamma} V_{\beta\gamma} \mp V_{\alpha\gamma} U_{\beta\gamma}) = 0,$$

respectively, and in the fermion case additionally $f_\alpha \equiv 0$. These transformations induce unitary transformations of the Fock space, if and only if the kernels $U_{\alpha\beta}$ and $V_{\alpha\beta}$ and the function f_α are quadratically summable (i.e. Lebesgue integrable) functions of the variables α and β, which in the thermodynamic limit is not true in all of the considered cases. (In (4.4, 6.15) and (4.36, 4.37) the kernels $U_{\alpha\beta}$ and $V_{\alpha\beta}$ contain the δ-function $\delta(\boldsymbol{p}_\alpha - \boldsymbol{p}_\beta)$ which may also appear in (6.18).) Of course, the above given class does by no means exhaust all canonical transformations of the canonical (anti-)commutation relations.

Analogous examples may be given for the case of a magnon spectrum in a magnetically ordered crystal. The most elementary case is the necessity of an (u,v)-transformation analogous to (4.36, 4.37) when introducing quasi-particles in an exchange ferromagnet and taking into account the dipole-dipole interaction between atomic magnetic moments [Holstein and Primakoff, 1940, Akhieser et al., 1967]. Magnetic transitions in materials with several sublattices may serve as an arsenal for demonstrating both the renormalization of bosonic spectra (Re $\omega_\nu(\boldsymbol{q})$ and Im $\omega_\nu(\boldsymbol{q})$) caused by quasi-particle interaction and the reconstruction of the vacuum caused, e.g., even by zero vibrations of magnons [Kaganov and Chubukov, 1987].

In this way, if the mean quasi-particle densities change, for instance due to a change of temperature or of pressure, the quasi-particle vacuum (corresponding to the cyclic representation vector) and the quasi-particle spectra also change. Often these are quantitative changes as expressed in (5.3) and obtained from the Green's functions (3.52) or (3.61) at non-zero temperature, with a possible temperature dependent linear transformation of the quasi-momenta $\boldsymbol{q}$. If, however, a boson mode thereby becomes soft (Re $\omega_\lambda(\boldsymbol{q}_c) \to 0$ at some wavevector $\boldsymbol{q}_c$), a new type of a Bose condensate will form. This includes the possibility of a 'condensate of the displacement field' ($\boldsymbol{R}_n$, $\boldsymbol{u}_{ns}$) resulting in a lattice transformation. If a fermion branch passes through the chemical potential, either a new sheet of the Fermi surface will appear or one of the sheets will disappear (in a degenerate case there is additionally the possibility of a change in the topological connection of the Fermi surface). In the case of superconductivity the fermionic spectrum reconstructs on the whole Fermi surface. In all these cases one has to do with qualitative reconstructions, that is with phase transitions.[37]

In the thermodynamic limit, these transitions are accompanied by the appearance of anomalous mean values which form the order parameters of the spontaneous symmetry breaking characterizing the transition. Technically, one enforces the symmetry breaking *before the thermodynamic limit* in the Hamiltonian by applying a corresponding external field the response to which generates the anomalous mean values. Then, *after the thermodynamic limit*, the limit of zero field is considered. In the case of a spontaneous symmetry breaking a non-zero anomalous

---

[37] A change of the topological connection of the Fermi surface implies a phase transition (of the order $2\frac{1}{2}$ in the nomenclature of Ehrenfest) strictly speaking only at zero temperature (e.g. as a function of pressure or chemical composition [Lifshits, 1960, Lifshits et al., 1973, Chap. 13]). At $T \neq 0$ such a change is accompanied with smoothed anomalies of physical quantities. To allow for a topological change of the Fermi surface either a self-intersecting isoenergetical surface must exist or a point in $\boldsymbol{p}$-space where a new sheet of the Fermi surface may appear (extremal point in the dispersion relation).

mean value survives at the end. This technical trick, proposed by N. N. Bogolubov since 1960 (Preprint 1451, JINR Dubna, 1963), goes under the name of *Bogolubov's quasi-means*. It provides the superselection of the order parameter which was discussed in Subsection 1.3.2 for the case of magnetic order (where the anomalous mean value or quasi-mean is the magnetization density).

6.4 Goldstone and Higgs Modes

Let us return for a moment to the considerations in connection with (6.3) to (6.7). Let the Hamiltonian of the system have a continuous symmetry group, but let its macroscopically homogeneous ground state be non-invariant with respect to this group, implying that the ground state level is infinitely degenerate (case of spontaneous symmetry breaking). Let the observable (6.6) discriminate these degenerate ground states, that is, if $|\Psi\rangle$ and $|\Psi'\rangle$ are two such states, then $\langle\Psi|\hat{b}|\Psi\rangle = b \neq b' = \langle\Psi'|\hat{b}|\Psi'\rangle$ (anomalous mean values). Consider a non-homogeneous excited state $|\Psi_V\rangle$ which inside the volume V equals $|\Psi'\rangle$, but in the remaining volume of the system is identical with $|\Psi\rangle$.[38] If the interaction in the system is short-ranged, the excitation energy E is obviously proportional to the surface of the volume V since the homogeneous states $|\Psi\rangle$ and $|\Psi'\rangle$ were degenerate:

$$E \sim V^{2/3}. \tag{6.19}$$

The state $|\Psi_V\rangle$ may be thought of being formed out of $|\Psi\rangle$ by exciting quasi-particles inside of V which have wavelengths $\lambda \lesssim V^{1/3}$. Their number should be proportional to V, and hence their energy $\varepsilon \sim E/V$. Together with (6.19) this gives

$$\varepsilon \sim V^{-1/3} \lesssim \lambda^{-1} \sim |\boldsymbol{p}|. \tag{6.20}$$

As $|\boldsymbol{p}| \to 0$, the quasi-particle energy ε goes to zero. The conclusion is that under the considered circumstances a spontaneous breaking of a continuous symmetry results in the appearance of a gapless (Goldstone) quasi-particle branch [Goldstone *et al.*, 1962, Nambu and Jona-Lasinio, 1961]. For instance in a crystal the translational invariance (shift of the crystal as a whole) is spontaneously broken. Here, the corresponding discriminating extensive observable $\hat{b}$ is constructed

---

[38]That means that all projections of $|\Psi_V\rangle$ onto states localized within V coincide with the corresponding projections of $|\Psi'\rangle$, and all projections onto states localized outside V coincide with those of $|\Psi\rangle$.

with the $\hat{u}_{ns}$ playing the role of $\hat{B}(x)$ of (6.6), and the Goldstone particles are the acoustic phonons. In a pure exchange ferromagnet[39] the role of $\hat{B}(x)$ is played by the spin density, and the Goldstone particles are acoustic magnons. In the superconducting state of the BCS model the gauge symmetry is spontaneously broken, but there is no Goldstone particle, since the model contains a long-range interaction producing the off-diagonal long-range order ($\langle|\hat{\psi}^\dagger\hat{\psi}^\dagger|\rangle \neq 0 \neq \langle|\hat{\psi}\hat{\psi}|\rangle$). If a local gauge symmetry is spontaneously broken, then instead of a Goldstone particle a longitudinal mode of the gauge field with finite non-zero energy, the Higgson, appears [Higgs, 1966]. An example for it is the plasmon in an electron liquid (the presence of a finite homogeneous electron density breaks the local gauge symmetry[40]).

If $\langle\hat{\psi}^\dagger\hat{\psi}\rangle = n \neq 0$ and $\hat{\psi} \to \hat{\psi}e^{i\theta/\hbar}$ where originally $\hat{\psi}$ had a constant (position independent) phase, then the charge current density is

$$j = \frac{q}{2m}\langle\hat{\psi}^\dagger\left(\frac{\hbar}{i}\overset{\leftrightarrow}{\nabla} - 2q\hat{A}\right)\hat{\psi}\rangle = \frac{nq}{m}(\nabla\theta - q A). \tag{6.21}$$

In a metal with no current flowing and no external magnetic field applied the scalar potential U is constant in space. A gauge can be used where $\theta = U = A = 0$. Consider now an excitation with $\delta j(t)$, $\delta A(t)$ in the 'wave gauge', where still $U = 0$ and hence $\theta = 0$ and

$$\delta j = -\frac{nq^2}{m}\delta A. \tag{6.22}$$

The continuity equation for the charge current is $q\partial\delta n/\partial t + \nabla \cdot \delta j = 0$. If it is once more differentiated with respect to time and if $\nabla \cdot (\partial A/\partial t) = -\nabla \cdot E = -4\pi q\delta n$ is observed which holds true in the chosen gauge (the last equality is Gauss's law), then (6.22) leads to

$$\frac{\partial^2 \delta n}{\partial t^2} = -\frac{4\pi q^2 n}{m}\delta n = -\omega_{\text{pl}}^2 \delta n, \qquad \omega_{\text{pl}} = \sqrt{\frac{4\pi q^2 n}{m}}. \tag{6.23}$$

---

[39]The dipole-dipole interaction in a magnetic material is long-ranged. A magnetic anisotropy (caused by spin-orbit coupling or by long-range electromagnetic coupling to the surface (shape of the sample) breaks already the symmetry in the Hamiltonian. Both cases lead to a gap in the magnon spectrum (absence of the Goldstone mode).

[40]The gauge field is the electromagnetic field in this case. This mechanism was discovered by P. W. Anderson in 1963 [Anderson, 1963] (cf. also [Anderson, 1958]) and served P. W. Higgs, F. Englert and R. Brout as the prototype for the corresponding theorem in particle physics [Higgs, 1966].

6.4 Goldstone and Higgs Modes

Since $n \neq 0$ breaks the gauge symmetry, instead of an acoustic Goldstone mode the plasmon appears at non-zero frequency ω_{pl}, the Higgson of a metal.

Consider now a superconductor with a homogeneous density, where $\boldsymbol{E} = 0$ holds even for a stationary $\boldsymbol{j} \neq 0$. Now, from $\nabla \cdot \boldsymbol{j} = 0$, even in the transverse gauge $\nabla \cdot \boldsymbol{A} = 0$ from (6.21) $\Delta \theta = 0$ follows, and hence $\theta = 0$ may be chosen. Then, Ampere's law $\Delta \boldsymbol{A} = -\mu_0 \boldsymbol{j}$ results in

$$\Delta \boldsymbol{A} = \frac{\mu_0 q^2 n}{m} \boldsymbol{A}. \tag{6.24}$$

Here, q is the pair charge and m is the pair mass; μ_0 is the vacuum permeability. The vector potential is screened with a screening length $\lambda = \sqrt{m/\mu_0 q^2 n}$, which is London's penetration depth, and a magnetic field cannot penetrate into a superconductor. Recalling that we assumed $\boldsymbol{j}$ stationary, $\Delta \boldsymbol{A} \sim \boldsymbol{j}$ and $\boldsymbol{A} = 0$ in the bulk of a superconductor (in a homogeneous superconducting state) again imply that zero-frequency electronic modes are absent there.

7 What Matter Consists of

Since ancient times, since man tried to understand nature, he tried to comprehend matter as consisting of elementary parts the interplay of which causes its properties. Empirically, we are given the various forms of matter with their properties. The elementary parts are mainly introduced to obtain a systematic and sufficiently simple characterization of general matter.

If a chemist is asked, what consists, say a piece of aluminum metal of, he will probably realize how he would sintesize it and answer: it consists of aluminum atoms, or even more likely: it consists of aluminum ions and of electrons. But maybe, recalling the well known experiment of Rutherford, he will say: it consists of aluminum nuclei and of electrons. A physicist will direct onto that piece of aluminum beams of various test particles, practically repeating the Rutherford experiment with that metallic crystal, and find that aluminum metal consists of phonons and Bloch electrons weakly interacting with each other. Using his experimental data, he may write down a Hamiltonian of the type (4.65). Restricted to this knowledge, he hardly would come to the conclusion that aluminum consists of nuclei and electrons interacting mutually with Coulomb forces. But being theoretically well educated he might find out that at low temperatures the Bloch electrons on the Fermi level may form Cooper pairs and that thereby the particle spectrum will be reconstructed with a gap opening up. In other materials he might also find that certain phonons may bind into biphonons [Pitaevskii, 1976], that charge and spin degrees of freedom may separate [Smirnov and Tsvelik, 2003, Bernevig *et al.*, 2001, and references therein], that fractionally charged particles appear [Laughlin, 1983, Heinonen, 1998], and so on.

Instead of extracting a Hamiltonian out of the data of scattering experiments, thermodynamic and kinetic data together with the connection between (3.35) and (4.65) and so on could be used. As is well known, exactly this was the way the physical ideas of metallic matter in the theories of P. Drude, A. Sommerfeld, F. Bloch, and others developed. (Recall that in the days of P. Drude the atomistic nature of condensed matter was not yet well recognized.)

If the physicist would have investigated the piece of aluminum at once at very low temperatures, he would directly have found that it consists of Bogolubov particles (hybrids of electrons and holes in the superconducting state) with a gap in their spectrum, phonons, possibly biphonons, and so on, interacting with each other, and he would have written down the corresponding Hamiltonian. On the other hand, by investigating the sample above the melting point, he would have arrived at the result that it consists of individual ions and an electronic Fermi liquid; above the boiling point he would have found basically neutral atoms, and, in the hot plasma state, nuclei and electrons.

Nowadays we have at hand technical-theoretical tools allowing us to start with nuclei, electrons and Coulomb forces (or even retarded electromagnetic forces in relativistic theories) or, as one is used to say, to start 'from the first principles' and, with moderate but sometimes also with astonishingly high accuracy, to establish the spectra of Bloch electrons and phonons. But without intermediately writing down the quasi-particle model Hamiltonian (4.65) we hardly would draw any conclusion on the nature of superconductivity.

7.1 The Hierarchy of Hamiltonians

So we find ourselves in a situation where we have to deal with a *hierarchy* of Hamiltonians of a solid each of which describes approximately the low-energy quasi-particle spectrum of the preceding one. 'Downwards' the hierarchy the abundance and complexity of diagonal terms in $\hat{H}_0$ of (4.65) increases and in the 'upward' direction the strengths of interactions in $\hat{H}_{\text{int}}$ grow. However, not at any level one could get rid of interaction because quasi-particles always decay (even in ideal infinite crystals). In the best case the quasi-particle damping decreases faster than its energy when approaching zero so that the condition (3.36) holds down to arbitraryly low excitation energies. As was already discussed, the lowest excitations may turn out to be stable altogether.

The presence of any structural disorder leads, however, to a damping which does not go to zero at all. (Therefore the term of a *pure sample* depends on the temperature of investigation and means that at a given temperature T in the energy range $|\varepsilon_\alpha - \mu| \approx k_B T$ the condition (3.36) is fulfilled.) In this situation of disorder the interpretation of the low-temperature properties in terms of quasi-particles runs into difficulties, and one is forced to have recourse to models with inner barriers hindering quasi-particle reactions (as e.g. a double-well potential [Anderson *et al.*, 1972]) and so on. The quasi-particle picture gets gradually lost. As any physical theory, it has its range of applicability, by no means comprising all of solid state physics.

In connection with the principally approximative character of the quasi-particle conception (see Section 2) the Hamiltonian of each level in the hierarchy is to some extent ambiguous and inaccurate, just for this reason it is called a *model Hamiltonian*. Its diagonal 'one-particle' part $\hat{H}_0$ at each level already *contains basically all the interaction of the preceding level*, resulting in the formation of the quasi-particles, and $\hat{H}_{\text{int}}$ contains only a more or less weak remainder interaction leading to the scattering of the quasi-particles on each other and leading to their reactions as well as to a possible reconstruction of the state and its quasi-particle spectra at lower temperatures. At each level a 'one-particle' mean-field theory may be used to estimate the mean effects of interaction on this level, or many-body techniques may be applied to penetrate into the details of the spectra, and eventually to pass over to the next lower level.

It is worthwhile to mention in this context, that the expression *one-particle approximation* has of course a definite meaning only in connection with the reference to which level it is applied, that is, with the reference to the Hamiltonian. As is, for instance, well known, the Hartree-Fock approximation applied to the Hamiltonian of nuclei, electrons, and their Coulomb interaction never yields a metallic state (because the logarithmic singularity in the Hartree-Fock self-energy operator caused by the long-range Coulomb interaction leads to an infinite Fermi velocity and hence to a vanishing density of states $D(\varepsilon_F)$). On the other hand, the same approximation when applied to the Hamiltonian of a metallic system of phonons and Bloch electrons with a short-range interaction (the long-range part of the 'original' Coulomb interaction being screened away in a special renormalization procedure; see (A.7) in Appendix I) not only gives the correct answer that the interaction does not destroy the metallic state but also allows for a qualitative investigation of fine-structures in the spectra appearing due to the presence of a Fermi surface. The anomalies of the well known Kohn-Migdal type, of the Taylor type [Taylor, 1963], and analogous anomalies in the electron spectrum [Kaganov, 1985, Eschrig and Kaganov, 1987] are found. But even at this level the Hartree-Fock approximation sometimes completely fails e.g. to explain qualitatively the spin structure of a magnetic metal [Fulde *et al.*, 1987].

7.2 'From the First Principles'

The term *from the first principles* has of course an analogous relative meaning and needs a reference of the level to which it is applied. To end this chapter, let us consider a bit more in detail the 'chemical' Hamiltonian of a non-relativistic

7.2 'From the First Principles'

ion-electron system from the angle of the just given discussion. The Hamiltonian is

$$\hat{H} = \sum_{pa} \hat{a}^\dagger_{pa} \frac{\boldsymbol{p}^2}{2M_a} \hat{a}_{pa} + \sum_{ps} \hat{c}^\dagger_{ps} \frac{\boldsymbol{p}^2}{2m} \hat{c}_{ps} +$$
$$+ (\hbar e)^2 \Bigg\{ \sum_{pa\, p'a'\, q} \hat{a}^\dagger_{p+q,a} \hat{a}^\dagger_{p'-q,a'} \frac{4\pi Z_a Z_{a'}}{q^2} \hat{a}_{p'a'} \hat{a}_{pa} +$$
$$+ \sum_{ps\, p's'\, q} \hat{c}^\dagger_{p+q,s} \hat{c}^\dagger_{p'-q,a'} \frac{4\pi}{q^2} \hat{c}_{p's'} \hat{c}_{ps} -$$
$$- \sum_{pa\, p's\, q} \hat{a}^\dagger_{p+q,a} \hat{c}^\dagger_{p'-q,s} \frac{4\pi Z_a}{q^2} \hat{c}_{p's} \hat{a}_{pa} \Bigg\}. \tag{7.1}$$

Here, $\hat{a}^\dagger_{pa}$ creates an ion of the sort a with mass M_a, valence Z_a, and with momentum $\boldsymbol{p}$. the operator $\hat{c}^\dagger_{ps}$ creates an electron with mass m and spin polarization s, and the summations over the momenta are to be understood in the sense

$$\sum_{\boldsymbol{p}} \equiv \frac{V}{(2\pi\hbar)^3} \int d^3 p \tag{7.2}$$

(where some small sphere around $\boldsymbol{q} = 0$ must be cut out in (7.1) the radius of which may go to zero only at the end of calculations meaning an Ewald-like treatment of the electrostatics [Ewald, 1921]). $\hat{H}$ has here the meaning of the total energy operator per unit volume.

First of all, this Hamiltonian is distinguished in the sense that it contains only particles which in the vacuum corresponding to the representation of (7.1) are *stable* (not posing questions here, e.g., about the stability of the proton) and the interaction between which has rigorously experimentally been established by two-particle scattering data (Rutherford experiment). Second, $\hat{H}$ commutes with the particle number operators of all sorts separately, so that from (7.1) the correctly defined self-adjoined Hamiltonian in any subspace of the Fock space characterized by fixed (finite) particle numbers of all sorts is readily obtained. This leads back to the usual Schrödinger equation of a system of N particles the solution of which in principle yields the energies and wave-functions of the ground state and of the exited states of the various chemical entities as atoms, molecules, radicals, and so on. Such a task is nowadays especially successfully solved using *density functional methods* [Eschrig, 2003, and references therein; see also Appendix A II]. With the help of the latter technique in recent years also the electron and ion density (the latter meaning the crystal structure) of various solids have been successfully calculated. True, practically all these calculations are up to now realized using the so-called *local-density approximation* the applicability of which is not justified but by its success. Since, however, the formal expressions for the needed functionals are taken from the well developed theory of the homogeneous electron liquid (a model which has strictly speaking no correspondence in nature but is sufficiently simple to be treated theoretically), those calculations do not imply any phenomenological parameter save for those contained in the Hamiltonian (7.1) (masses, charges, $\hbar$).

Moreover, within these approaches the accuracy of the Hamiltonian (7.1) may be improved in several respects: the kinetic energy expression may be replaced by a relativistic one resulting in the Dirac-Coulomb Hamiltonian and the Breit interaction may be added accounting for retardation and spin dependence (dipole-dipole interaction) of the photon exchange between electrons and between the electron and the nucleus [Strange, 1998]. These corrections may also be considered in a simpler perturbative way by adding instead a spin-orbit coupling term, a hyperfine term and so on.

Using still another local density approximation for the self-energy operators $\hat{\Sigma}$ of Bloch electrons it becomes possible, from (7.1) to calculate the one-particle part of the quasi-particle Hamiltonian of phonons and Bloch electrons

$$\hat{H}_0^q = \sum_{q\lambda} \hat{b}_{q\lambda}^\dagger \, \hbar\omega_\lambda(q) \, \hat{b}_{q\lambda} + \sum_{p\nu} \hat{d}_{p\nu}^\dagger \, \varepsilon_\nu(p) \, \hat{d}_{p\nu}, \qquad (7.3)$$

which agrees well with experimental data for the excitation spectra of solids. This part of the way is, however, even more phenomenological. Indeed, for a finite non-zero particle density in the thermodynamic limit the Hamiltonian (7.1) is ill-defined, and on the way to the equation (5.23) for the pole of the Green function, in which $\hat{\Sigma}$ enters, all kinds of renormalizations (partial summations) are necessary (Appendix I), none of which is performed rigorously. Furthermore. little is known about the possibility to extract from (7.1) also the *interaction* Hamiltonian $\hat{H}_{\text{int}}^q$ of the quasi-particles entering (7.3) (see Section 3). However, after all, establishing (7.3) 'from the first principles' must really be regarded an enormous success of theory, at least in view of its enormous predictive power.

8 Epilogue: The Unifying Picture of Physics

The picture sketched in the last two chapters is very much reminiscent of the modern picture in particle physics, that is of the physics of particles moving in the Poincaré vacuum. In fact the high (Poincaré) symmetry of the vacuum in this case imposes strong constraints on the possible shape of the particle spectra reducing them to

$$\varepsilon_\nu(\boldsymbol{p}) = c\sqrt{m_\nu^2 c^2 + \boldsymbol{p}^2}, \tag{8.1}$$

where only the particle masses m_ν remain as parameters (including) the particle lifetimes in the vacuum as imaginary parts of their masses). For the rest, the arguments may be continued without alterations: Investigating matter at higher energies one finds that nuclei consist of nucleons interacting with each other by pion exchange, the latter, in the stationary state, forming Yukawa forces. At even higher energies it turns out that nucleons and pions consist of quarks interacting by gluon exchange and that even letpons exchange, besides photons, heavy gauge bosons of the weak interaction. The vacuum thereby not only influences the dynamical parameters of the particles so that due to a vacuum polarization around a given particle the 'bare' masses and charges do not reveal themselves in experiment but instead 'dressed' ones, renormalized due to interactions, just as in a metallic crystal Bloch electrons manifest with a screened short-range interaction instead of usual electrons interacting with Coulomb forces. Moreover, in modern understanding [Gaillard et al., 1999] the vacuum itself is determined by the presence of fields, and reversely it determines through the condensate of the Goldstone-Higgs field the masses and charges of particles. Thereby the condensate of the Goldstone-Higgs field may change in time [Linde, 1985, Albrecht and Brandenberger, 1985] according to changes of the particle densities in the universe.

Today one speaks already of quarks and leptons as consisting of 'preons', and the end of the way 'upwards' in the hierarchy seems not yet visible. It is, however, likely that somewhere on this way the particle picture gets also lost.[41] Up to now,

---

[41] The last development considers elementary particles as low-energy (on a scale of 10^{28}

from experiment not too much is known with certainty about quarks and most leptons and nothing about their 'preons' so that climbing up the hierarchy only by theoretical reasoning is very speculative.

The picture layed out leads to the following qualified scheme:

?
↑
Superstrings
GUT (grand unification theory)
SUSY (supersymmetry)
QCD (quantum chromodynamics)
Nuclear chemistry (nucleons, pions,...)
.
Chemistry (nuclei, electrons,...)
Solid state physics (phonons, Bloch electrons, magnons...)
Low temperature physics (Bogolubov electrons, biphonons,
        holons, spinons, phasons, ...)
↓
?

Each level in this scheme has its energy range which via the condition (3.34) is connected with the strengths of interaction of the particles entering the level. In connection with the fact that because of the non-ideal structure of any real matter the interaction at low excitation energies always remains finite, the particle picture gets lost in certain cases in the limit of low energies. In the opposite direction, due to a rapid growth of interaction with increasing energies, it gets very likely lost too.

In the above sequence, a dot was put between the level of chemistry and that of nuclear chemistry. This dot marks a larger jump in the hierarchy of energy scales (six orders of magnitude from eV to MeV) which in fact provides the deeper reason for the possibility to speak of an approach 'from the first principles' in Quantum

---

eV) quasi-stationary excitations of closed strings (on a scale of 10^{-35} m) in a high-dimensional unitary space or even more complex situations as strings spanned between 'branes' [Schwarz and Seiberg, 1999, and references therein]. In a certain sense the particle conception already might be in doubt in the case of quarks which do not exist as free particles in space. They are, however, seen like free particles inside of nuclei ('partons'), and, in the understanding of this paper, this appearance legitimates them without reserve as particles: Phonons do also not exist outside of condensed matter, nevertheless they are generally recognized as being particles (although quasi-...).

8 Epilogue: The Unifying Picture of Physics

Chemistry (Hamiltonian (7.1) possibly with relativistic corrections; the neglected light neutral leptons, the neutrinos, do practically not interact with chemical matter; the interaction bosons $W^\pm$ and Z have heavy masses of the order $mc^2 \sim 100$ GeV). If the Higgson will eventually be found experimentally (expected in the range from 80 GeV to 130 GeV) and the supersymmetric particles will not be found (expected in the TeV range), then a colossal jump of fourteen orders of magnitude would result from QCD below 10^2 GeV to GUT above 10^{16} GeV with no new particles in the then anticipated so-called 'particle desert' in between. Then, one would with every justification speak of the QCD as a closed theory, although it contains a lot of phenomenological parameters, containing in their relations and energy dependences already the hallmark of a higher theory.

As we see, there is no principal difference between 'particles' and 'quasi-particles'. Quasi-particles are particles in a generalized understanding, moving in the reduced symmetry of condensed matter, and the common particles are quasi-particles moving in a Poincaré vacuum.

Thus, before us, an unified and exceptionally beautiful picture appears: a unique conception applies with equal success to phenomena from the micro Kelvin range up to the temperatures of the 'big bang' of the universe. It is just this feature which proves the vivid power of the physical approach to the exploration of nature.

Appendices

A I Self-energy operator

For a non-interacting system described by the Hamiltonian $\hat{H}_0$ of (4.65), using (3.21) the Green function may be written as[42]

$$\hat{G}_0(\omega) = \left[\omega + \mu - \hat{H}_0\right]^{-1}. \tag{A.1}$$

The subscript '0' indicates the absence of interaction. Defining

$$\hat{\Sigma} \equiv \hat{G}_0^{-1} - \hat{G}^{-1}, \tag{A.2}$$

we find the perturbation series

$$\begin{aligned}\hat{G} &= \left[\hat{G}_0^{-1} - \hat{\Sigma}\right]^{-1} = \hat{G}_0 + \hat{G}_0\hat{\Sigma}\hat{G}_0 + \hat{G}_0\hat{\Sigma}\hat{G}_0\hat{\Sigma}\hat{G}_0 + \cdots = \\ &= \hat{G}_0 + \hat{G}_0\hat{\Sigma}\hat{G}\end{aligned} \tag{A.3}$$

for the Green's function $\hat{G}$ of the interacting system described by the Hamiltonian $\hat{H}$ of (4.65). The expression (A.3) may be recast as

$$\left[\hat{G}_0^{-1} - \hat{\Sigma}\right]\hat{G} = \hat{I}, \tag{A.4}$$

which in view of (A.1) is identical with (5.23) or (5.36). The actual connection between $\hat{\Sigma}$ and $\hat{H}_{\text{int}}$ is not used in any detail in the present context, it may be found in textbooks [Abrikosov et al., 1975, Landau and Lifshits, 1980b].

Graphically, the series (A.3) may be represented as

$$\text{(A.5)}$$

If one introduces a wavy line for the two-particle interaction matrix element of $\hat{H}_{\text{int}}$, then $\hat{\Sigma}$ appears as the series

$$\text{(A.6)}$$

---

[42]In the Appendices we put $\hbar = 1$.

8.1 Density functional theory

This series may be partially summed up by introducing an effective interaction

$$\text{(diagram)} \tag{A.7}$$

Then,

$$\text{(diagram)} \tag{A.8}$$

Neglecting here all higher order terms, the following closed system of equations is obtained:

$$\hat{G} = \hat{G}_0(1 + \hat{\Sigma}\hat{G}), \quad \hat{\Sigma} \approx \hat{G}\hat{W}, \quad \hat{W} \equiv \hat{H}_{\text{int}}(1 + \hat{G}\hat{G}\hat{W}). \tag{A.9}$$

This is called the GW approximation for $\hat{\Sigma}$ [Hedin and Lundqvist, 1971]. E.g. from the graphs explicitly shown in (A.6) only the last one is not covered by (A.9). Nowadays these equations may be solved numerically even for inhomogeneous electron systems by using powerful computers [Hybertsen and Louie, 1985], provided the ground state electron density of the inhomogeneous system has independently been obtained.

A II Density functional theory

In the subspace of fixed particle numbers the expression (7.1) reduces to the Schrödinger Hamiltonian

$$\hat{H} = \sum_a \frac{\nabla_a^2}{2M_a} + \sum_{aa'} \frac{Z_a Z_{a'} e^2}{|\boldsymbol{R}_a - \boldsymbol{R}_{a'}|} + \hat{H}_{\text{el}}, \tag{A.10}$$

$$\hat{H}_{\text{el}} = \sum_s \frac{\nabla_s^2}{2m} + \sum_{ss'} \frac{e^2}{|\boldsymbol{r}_s - \boldsymbol{r}_{s'}|} + \sum_{sa} \frac{Z_a e^2}{|\boldsymbol{R}_a - \boldsymbol{r}_s|} \equiv \hat{T} + \hat{W} + \hat{V}, \tag{A.11}$$

where here the subscripts a and s, counting the nuclei and electrons, respectively, are used in a different meaning compared to (7.1). The presence of a small parameter $(m/M)^{1/2} \approx 10^{-3/2} \cdots 10^{-5/2}$ allows for the neglect of the ionic motion in the electron problem (adiabatic approximation):

$$\hat{H}_{\text{el}} \Psi(\boldsymbol{r}_s; \boldsymbol{R}_a) = E(\boldsymbol{R}_a) \Psi(\boldsymbol{r}_s; \boldsymbol{R}_a). \tag{A.12}$$

Here, the $\boldsymbol{R}_a$ are assumed fixed, and Ψ is the electronic ground state for a given configuration of fixed ions at $\boldsymbol{R}_a$. Using the result of (A.12), the Hamiltonian (A.10) reduces to the adiabatic Hamiltonian

$$\hat{H}_{\mathrm{ad}} = \sum_a \frac{\nabla_a^2}{2M_a} + U(\boldsymbol{R}_a), \qquad (A.13)$$

where the effective adiabatic potential

$$U(\boldsymbol{R}_a) = \sum_{aa'} \frac{Z_a Z_{a'} e^2}{|\boldsymbol{R}_a - \boldsymbol{R}_{a'}|} + E(\boldsymbol{R}_a) \qquad (A.14)$$

may be expanded in a power series with respect to the ionic displacements $\boldsymbol{u}_a = \boldsymbol{R}_a - \boldsymbol{R}_a^0$ from their equilibrium positions $\boldsymbol{R}_a^0$:

$$U(\boldsymbol{R}_a) \approx U_0 + \frac{1}{2} \sum_{aa'} \boldsymbol{u}_a \cdot \boldsymbol{\Phi}_{aa'} \cdot \boldsymbol{u}_{a'}. \qquad (A.15)$$

The diagonalization of the rescaled force constant matrix $\boldsymbol{\Phi}_{aa'}/\sqrt{M_a M_{a'}}$, solves the problem of the normal modes of ionic vibrations (phonon spectra in a solid [Maradudin, 1974]).

It remains to solve the problem (A.12). Instead of indicating the dependence on the $\boldsymbol{R}_a$ the ground state Ψ of the N-electron system may be considered as a functional of the external potential

$$v(\boldsymbol{r}) = \sum_a \frac{Z_a e^2}{|\boldsymbol{r} - \boldsymbol{R}_a|}. \qquad (A.16)$$

This functional dependence of $\Psi[v]$ is transferred to all ground state properties, e.g. the energy

$$E[v] = \langle \Psi[v]|\hat{H}[v, N]|\Psi[v]\rangle = \inf_v \left\{ \langle \Psi|\hat{H}[v, N]|\Psi\rangle \,\Big|\, \langle \Psi|\Psi\rangle = 1 \right\}, \qquad (A.17)$$

where $\hat{H}[v, N]$ is written instead of $\hat{H}_{\mathrm{el}}$ in order to indicate the dependence on v and the electron number N explicitly. We switch again to a grand canonical description and replace (A.17) by the infimum over a general mixed state (2.17),

$$E[v, N] = \inf_{\hat{\rho}} \left\{ \mathrm{tr}(\hat{\rho}\hat{H}[v]) \,\Big|\, \mathrm{tr}(\hat{\rho}\hat{N}) = N \right\}, \qquad (A.18)$$

A II Density functional theory

where, however, $\hat{H}[v]$ does not contain the term $\mu\hat{N}$. It is now rather easy to show that $E[v, N]$ is convex in N and concave in v, and that its twofold Legendre transform is (cf. for what follows [Eschrig, 2003, Chapter 6]; in contrast to most publications on density functional theory it is all rigorous)

$$H[n] = \inf_N \sup_v \{E[v, N] - (v|n)\}, \tag{A.19}$$

where a scalar product $(v|n) = \sum_s \int d^3r \, v_{ss'}(\boldsymbol{r}) n_{s's}(\boldsymbol{r})$ was introduced and in addition to the potential (A.16) a spin dependent part was allowed for.

If the universal density functional $H[n]$ would be known, then $E[v, N]$ would follow from the Legendre back transformation which is the variational principle by Hohenberg and Kohn [Hohenberg and Kohn, 1964]

$$E[v, N] = \inf_n \{H[n] + (v|n) \mid (1|n) = N\}. \tag{A.20}$$

Here, $(1|n)$ simply means the integral over the electron density $n(\boldsymbol{r}) = \sum_s n_{ss}(\boldsymbol{r})$. However, $H[n]$ is of course unknown.

Nevertheless, with an ingenious trick by Kohn and Sham [Kohn and Sham, 1965], one can go ahead quite far. The electron density is expressed through new variational quantities: the Kohn-Sham orbitals $\phi_i(\boldsymbol{r}, s)$ and their occupation numbers n_i,

$$n_{ss'}(\boldsymbol{r}) = \sum_i \phi_i(\boldsymbol{r}, s) n_i \phi_i^*(\boldsymbol{r}, s'), \quad \langle \phi_i | \phi_j \rangle = \delta_{ij}, \quad 0 \leq n_i \leq 1, \quad \sum_i n_i = N. \tag{A.21}$$

The unknown density functional is then represented as

$$H[n] = K[n] + L[n], \tag{A.22}$$

$$K[n] = \min_{\phi_i, n_i} \left\{ k[\phi_i, n_i] \,\Big|\, \sum_i \phi_i n_i \phi_i^* = n, \, 0 \leq n_i \leq 1, \, \langle \phi_i | \phi_j \rangle = \delta_{ij} \right\}, \tag{A.23}$$

with a suitably chosen orbital functional $k[\phi_i, n_i]$ and a remaining density functional $L[n]$. This casts the variational principle (A.20) into the variational principle by Kohn and Sham:

$$E[v, N] = \inf_{\phi_i, n_i} \left\{ k[\phi_i, n_i] + L\left[\sum_i \phi_i n_i \phi_i^*\right] + \sum_i \langle \phi_i | v | \phi_i \rangle \,\Big|\right.$$
$$\left. 0 \leq n_i \leq 1, \, \sum_i n_i = N, \, \langle \phi_i | \phi_j \rangle = \delta_{ij} \right\}. \tag{A.24}$$

Variation of the ϕ_i yields the Kohn-Sham equation

$$(\hat{k} + v + v_L)\phi_i = \phi_i \varepsilon_i, \qquad \hat{k}\phi_i = \frac{1}{n_i}\frac{\delta k}{\delta \phi_i^*}, \qquad v_L = \frac{\delta L}{\delta n} \qquad (A.25)$$

where the functional derivatives of k and L figure. The variation of the n_i results in the so called aufbau principle, where an ascending order

$$\varepsilon_1 \leq \varepsilon_2 \leq \cdots \qquad (A.26)$$

has to be adopted:

$$n_i = 1 \text{ for } \varepsilon_i < \varepsilon_N, \ 0 \leq n_i \leq 1 \text{ for } \varepsilon_i = \varepsilon_N, \ n_i = 0 \text{ for } \varepsilon_i > \varepsilon_N. \qquad (A.27)$$

So far goes the rigorous density functional theory.

Since $H[n]$ is unknown, the functionals $k[\phi_i, n_i]$ and $L[n]$ have to be modeled, and here starts the *model density functional theory* which compares to the use of model Hamiltonians in many-body theory and is overall astonishingly successful.

Of course, k may be chosen rather arbitrarily, which then defines the rigorous L (L is a density functional because K was defined as a density functional via the constrained minimum). The most frequent (but by far not the only) choice of k is

$$k[\phi_i, n_i] = \sum_i n_i \langle \phi_i | \hat{t} | \phi_i \rangle + \frac{1}{2}\sum_{ij} n_i n_j \langle \phi_i \phi_j' | |\boldsymbol{r} - \boldsymbol{r}'|^{-1} | \phi_i \phi_j' \rangle, \qquad (A.28)$$

where $\hat{t}$ can be the non-relativistic or the relativistic kinetic energy operator, depending on whether one is aiming at a Kohn-Sham or a Dirac-Kohn-Sham equation. Hence, the first term of k is just the kinetic energy of a non-interacting system. Observe, that k is this energy, not K, which is a constrained minimum of k. In the second, the Hartree term, the items $i = j$ (self-interaction) are included. For $L[n]$ one choses the *local density approximation*

$$L[n] = E_{\text{XC}}[n] = \int d^3r \, n(\boldsymbol{r}) \epsilon_{\text{XC}}[n_{ss'}(\boldsymbol{r})], \qquad (A.29)$$

where $\epsilon_{\text{XC}}[n_{ss'}(\boldsymbol{r})]$ is the exchange and correlation energy per particle of a homogeneous electron liquid with homogeneous density $n_{ss'} = \text{const.}$ equal to $n_{ss'}(\boldsymbol{r})$. It is known with high accuracy from quantum Monte Carlo calculations in combination with high density perturbation expansion. The ground state energy is then obtained as

$$E[v, N] = \sum_i \varepsilon_i - E_{\text{H}}[n] - \sum_{ss'} \int d^3r \, n_{ss'} v_{L,s's} + E_{\text{XC}}[n], \qquad (A.30)$$

A II Density functional theory

where $E_H[n]$ is the Hartree energy including the self interaction (classical electrostatic energy of a density $n(\boldsymbol{r})$). Various generalizations improving the local density approximation in chosen cases are in use.

This density functional theory yields the ground state energy and the ground state spin density of the electrons. Via (A.14) it yields also the structural energy of a solid and the adiabatic phonon spectra. Analogously, it also yields adiabatic magnons (they often make sense although no mass ratio is involved) and static responses to external fields. The Kohn-Sham orbital energies are often taken as approximate electronic quasi-particle spectra, which can be justified in many cases but may (depending on the used functionals) fail completely in others.

Bibliography

[Abrikosov, 1988] A. A. Abrikosov, *Fundamentals of the Theory of Metals* (North-Holland, Amsterdam, 1988).

[Abrikosov et al., 1975] A. A. Abrikosov, L. P. Gorkov, and I. E. Dzyaloshinski, *Methods of Quantum Field Theory in Statistical Physics* (Dover Publications, Inc., New York, 1975).

[Akhieser et al., 1967] A. I. Akhieser, V. G. Bar'yakhtar, and S. V. Peletminskii, *Spin Waves* (Interscience, New York, 1967).

[Albrecht and Brandenberger, 1985] A. Albrecht and R. H. Brandenberger, "Realization of New Inflation", Phys. Rev. **D31**, 1225–1231 (1985).

[Ambegaokar, 1961] V. Ambegaokar, "Electronic Properties of Insulators. II", Phys. Rev. **121**, 91–103 (1961).

[Anderson, 1958] P. W. Anderson, "Random-Phase Approximation in the Theory of Superconductivity", Phys. Rev. **112**, 1900–1916 (1958).

[Anderson, 1963] P. W. Anderson, "Plasmons, Gauge Invariance, and Mass", Phys. Rev. **130**, 439–442 (1963).

[Anderson, 1997] P. W. Anderson, *Concepts in Solids* (World Scientific, Singapore, 1997), first edition: Benjamin, New York 1963.

[Anderson et al., 1972] P. W. Anderson, B. I. Halperin, and C. M. Varma, "Anomalous Low-Temperature Thermal Properties of Glasses and Spin Glasses", Phil. Mag. **25**, 1–9 (1972).

[Ashcroft and Mermin, 1976] N. W. Ashcroft and N. D. Mermin, *Solid State Physics* (Rinehart and Winston, New York, 1976).

[Baranovskii et al., 1979] S. D. Baranovskii, A. L. Efros, B. L. Gelmont, and B. I. Shklovskii, "Coulomb Gap in Disordered Systems: Computer Simulation", J. Phys. C: Solid St. Phys. **12**, 1023–1034 (1979).

[Berezin, 1965] F. A. Berezin, *The Method of Second Quantization* (Academic Press, New York, 1965).

Bibliogaphy

[Bernevig et al., 2001] B. A. Bernevig, D. Giuliano, and R. B. Laughlin, "Spinon-Holon Attraction in the Supersymmetric $t - J$ Model with $1/r^2$ Interaction", Phys. Rev. Lett. **87**, 177206-1-4 (2001).

[Bogolubov, 1947] N. N. Bogolubov, "On the Theory of Superfluidity", J. Phys. (USSR) **11**, 23–32 (1947).

[Bogolubov, 1958] N. N. Bogolubov, "A New Method in the Theory of Superconductivity. I", Sov. Phys. JETP **7**, 41–46 (1958).

[Bogolubov and Shirkov, 1959] N. N. Bogolubov and D. V. Shirkov, *Introduction to the Theory of Quantized Fields* (Interscience, New York, 1959).

[Bohm, 1993] A. Bohm, *Quantum Mechanics: Foundations and Applications* (Springer-Verlag, Berlin, 1993).

[Brandt and Moshkalov, 1984] N. B. Brandt and V. V. Moshkalov, "Concentrated Kondo Systems", Adv. Phys. **33**, 373–467 (1984).

[Bratteli and Robinson, 1987] O. Bratteli and D. W. Robinson, *Operator Algebras and Quantum Statistical Mechanics, 1 and 2* (Springer-Verlag, New York, 1981, 1987).

[Brovman and Kagan, 1974] E. G. Brovman and Y. M. Kagan, "Phonons in Non-Transition Metals", in *Dynamical Properties of Solids*, edited by G. K. Horton and A. A. Maradudin (North-Holland, Amsterdam, 1974), Chap. 4, pp. 191–300.

[Bychkov and Gor'kov, 1961] Y. A. Bychkov and L. P. Gor'kov, "Quantum Oscillations of the Thermodynamic Quantities of a Metal in a Magnetic Field According to the Fermi-Liquid Model", Sov. Phys. JETP **14**, 1132–1140 (1961).

[Castellani et al., 1987] C. Castellani, G. Kottliar, and P. A. Lee, "Fermi-Liquid Theory of Interacting Disordered Systems and the Scaling Theory of the Metal-Insulator Transition", Phys. Rev. Lett. **59**, 323–326 (1987).

[Cohen and Heine, 1970] M. L. Cohen and V. Heine, in *The Fitting of Pseudopotentials to Experimental Data and Their Subsequent Application*, Vol. 24 of *Solid State Physics*, edited by H. Ehrenreich, F. Seitz, and D. Turnball (Academic Press, New York and London, 1970), pp. 37–248.

[Cook, 1953] J. M. Cook, "The Mathematics of Second Quantisation", Trans. Am. Math. Soc. **74**, 224–245 (1953).

[Davies et al., 1984] J. H. Davies, P. A. Lee, and T. M. Rice, "Properties of the Electron Glass", Phys. Rev. **B29**, 4260–4271 (1984).

[Dederichs et al., 1984] P. H. Dederichs, S. Blügel, R. Zeller, and H. Akai, "Ground States of Constrained Systems: Application to Cerium Impurities", Phys. Rev. Lett. **53**, 2512–2515 (1984).

[Deinzer et al., 2004] G. Deinzer, M. Schmitt, A. P. Mayer, and D. Strauch, "Intrinsic Lifetimes and Anharmonic Frequency Shifts of Long-Wavelength Optical Phonons in Polar Crystals", Phys. Rev. **B69**, 014304-1-7 (2004).

[Deppeler and Millis, 2002] A. Deppeler and A. J. Millis, "Electron-Phonon Interactions in Correlated Systems: Adiabatic Expansion of the Dynamical Mean-Field Theory", Phys. Rev **B65**, 100301-1-4 (2002).

[DeWitt, 1992] B. DeWitt, *Supermanifolds* (Cambridge University Press, Cambridge, 1992).

[Doğan and Marsiglio, 2003] F. Doğan and F. Marsiglio, "Self-Consistent Modification of the Electron Density of States Due to Electron-Phonon Coupling in Metals", Phys. Rev. **B68**, 165102-1-7 (2003).

[Eckert and Youngblood, 1986] J. Eckert and R. Youngblood, "Lattice Dynamics and Phonon Line Shapes in ^{36}Ar at High Temperatures", Phys. Rev. **B34**, 2770–2776 (1986).

[Efetov, 1997] K. Efetov, *Supersymmetry in Disorder and Chaos* (Cambridge University Press, Cambridge, U.K., 1997).

[Efros and Shklovskii, 1985] A. L. Efros and B. I. Shklovskii, "Coulomb Interaction in Disordered Systems with Localized Electronic States", in *Electron-Electron Interaction in Disordered Systems*, edited by A. L. Efros and M. Pollak (North-Holland, Amsterdam, 1985), pp. 409–482.

[Emch, 1972] G. G. Emch, *Algebraic Methods in Statistical Mechanics and Quantum Field Theory* (Wiley-Interscience, New York, 1972).

[Eschrig, 2003] H. Eschrig, *The Fundamentals of Density Functional Theory* (Edition am Gutenbergplatz, Leipzig, 2003).

[Eschrig and Kaganov, 1987] H. Eschrig and M. I. Kaganov, "Quasi-Particle Interaction and Anomalies in the Excitation Spectra of Metals", in *Physics of Phonons*, Vol. 285 of *Lecture Notes in Physics*, edited by T. Paszkiewicz (Springer, Heidelberg, 1987), pp. 334–347.

[Ewald, 1921] P. P. Ewald, "Die Berechnung Optischer und Elektrostatischer Gitterpotentiale", Ann. Physik (Leipzig) **64**, 253–287 (1921).

[Fetter and Walecka, 1971] A. L. Fetter and J. D. Walecka, *Quantum Theory of Many-Particle Systems* (McGraw-Hill Book Company, New York, 1971).

[Ford, 1975] J. Ford, "The Statistical Mechanics of Classical Analytical Dynamics", in *Fundamental Problems in Statistical Mechanics, Vol. 3*, edited by E. D. G. Cohen (North-Holland, Amsterdam, 1975).

[Fradkin, 1991] E. Fradkin, *Field Theories of Condensed Matter Systems* (Addison-Wesley Publishing Company, Redwood City, 1991).

[Fulde, 1995] P. Fulde, *Electron Correlations in Molecules and Solids* (Springer-Verlag, Berlin, 1995).

[Fulde et al., 1987] P. Fulde, Y. Kakehashi, and G. Stollhoff, "Electron Correlations in Transition Metals", in *Metallic Magnetism*, Vol. 42 of *Topics in Current Physics*, edited by H. Capellmann (Springer, Heidelberg, 1987), pp. 159–206.

[Gaillard et al., 1999] M. K. Gaillard, P. D. Grannis, and F. J. Sciulli, "The Standard Model of Particle Physics", Rev. Mod. Phys. **71**, S96–S111 (1999).

[Galitskii and Migdal, 1958] V. M. Galitskii and A. B. Migdal, "Application of Quantum Field Theory Methods to the Many Body Problem", Sov. Phys. JETP **7**, 96–104 (1958).

[Gelfand and Naimark, 1943] I. Gelfand and M. A. Naimark, "On the Imbedding of Normed Rings into the Ring of Operators in Hilbert Space", Math. Sborn., N. S. **12**, 197–217 (1943).

[Glauber, 1963] R. Glauber, "Coherent and Incoherent States of the Radiation Field", Phys. Rev. **131**, 2766–2788 (1963).

[Goldstone et al., 1962] J. Goldstone, A. Salam, and S. Weinberg, "Broken Symmetries", Phys. Rev. **127**, 965–970 (1962).

[Grib et al., 1971] A. A. Grib, E. V. Damaskinskii, and V. M. Maximov, "The Problem of Symmetry Breaking and Invariance of the Vacuum in Quantum Field Theory", Sov. Phys. Usp. **13**, 798–815 (1971).

[Gunnarsson et al., 1989] O. Gunnarsson, O. K. Andersen, O. Jepsen, and J. Zaanen, "Density-Functional Calculation of the Parameters in the Anderson Model: Application to Mn in CdTe", Phys. Rev. **B39**, 1708–1722 (1989).

[Gurevich and Parashkin, 1982] V. L. Gurevich and D. A. Parashkin, "Quantum Theory of Acoustic and Electromagnetic Nonresonant Absorption in Glasses", Sov. Phys. JETP **56**, 1334–1342 (1982).

[Guyer, 1969] R. Guyer, in *The Physics of Quantum Crystals*, Vol. 23 of *Solid State Physics*, edited by H. Ehrenreich, F. Seitz, and D. Turnball (Academic Press, New York and London, 1969), pp. 413–499.

[Haag, 1962] R. Haag, "The Mathematical Structure of the Bardeen-Cooper-Schrieffer Model", Nouvo Cimento **25**, 287–299 (1962).

[Haag, 1993] R. Haag, *Local Quantum Physics* (Springer-Verlag, Berlin, 1993).

[Haag et al., 1967] R. Haag, N. M. Hugenholtz, and M. Winnink, "On the Equilibrium States in Statistical Mechanics", Commun. Math. Phys. **5**, 215–236 (1967).

[Haag and Kastler, 1964] R. Haag and D. Kastler, "An Algebraic Approach to Quantum Field Theory", J. Math. Phys. **5**, 848–861 (1964).

[Hedin and Lundqvist, 1971] L. Hedin and B. I. Lundqvist, "Explicit Local Exchange and Correlation Potentials", J. Phys. C: Solid St. Phys. **4**, 2064–2083 (1971).

[Heinonen, 1998] *Composite Fermions*, edited by O. Heinonen (World Scientific, Singapore, 1998).

[Hewson, 1993] A. C. Hewson, *The Kondo Problem to Heavy Fermions* (Cambridge University Press, Cambridge, U.K., 1993).

[Higgs, 1966] P. W. Higgs, "Spontaneous Symmetry Breakdown without Massless Bosons", Phys. Rev. **145**, 1156–1163 (1966).

[Hohenberg and Kohn, 1964] P. Hohenberg and W. Kohn, "Inhomogeneous Electron Gas", Phys. Rev. **136**, B864–B871 (1964).

[Holstein and Primakoff, 1940] T. Holstein and H. Primakoff, "Field Dependence of the Intrinsic Domain Magnetization of a Ferromagnet", Phys. Rev. **58**, 1098–1113 (1940).

[Horton and Cowley, 1987] G. K. Horton and E. R. Cowley, "Lattice Dynamics at High Temperature", in *Physics of Phonons*, edited by T. Paszkiewicz (Springer-Verlag, Heidelberg, 1987), pp. 50–80.

[Hybertsen and Louie, 1985] M. S. Hybertsen and S. G. Louie, "Electron Correlation and the Band Gap in Ionic Crystals", Phys. Rev. **B32**, 7005–7008 (1985).

[Itzykson and Zuber, 1980] C. Itzykson and J.-B. Zuber, *Quantum Field Theory* (McGraw-Hill Book Company, New York, 1980).

[Kadanoff and Baym, 1989] L. P. Kadanoff and G. Baym, *Quantum Statistical Mechanics* (Addison-Wesley Publ. Co. Inc., New York, 1989).

[Kaganov, 1985] M. I. Kaganov, "Energy Spectrum of a Metal and its Singularities", Sov. Phys. Usp. **28**, 257–268 (1985).

[Kaganov and Chubukov, 1987] M. I. Kaganov and A. V. Chubukov, "Interacting Magnons", Sov. Phys. Usp. **30**, 1015–1040 (1987).

[Kaganov and Liftshits, 1980] M. I. Kaganov and I. M. Liftshits, *Quasi-Particles: Ideas and Principles of Solid State Physics* (Mir Publishers, Moscow, London, 1980).

[Kaganov and Lisovskaya, 1981] M. I. Kaganov and T. Y. Lisovskaya, "Effect of Electron-Phonon Interaction on the Electron Spectrum in a Normal Metal", Sov. Phys. JETP **53**, 1280–1283 (1981).

[Khalatnikov, 1985] I. M. Khalatnikov, *An Introduction to the Theory of Superfluidity* (Addison-Wesley Publ. Co., Inc., New York, 1985).

[Khomskii, 1979] D. I. Khomskii, "The Problem of Intermediate Valency", Sov. Phys. Usp. **22**, 879–903 (1979).

[Kohn, 1958] W. Kohn, "Interaction of Charged Particles in a Dielectric", Phys. Rev. **110**, 857–864 (1958).

[Kohn, 1959] W. Kohn, "Image of the Fermi Surface in the Vibrational Spectrum of a Metal", Phys. Rev. Lett. **3**, 393–340 (1959).

[Kohn and Sham, 1965] W. Kohn and L. J. Sham, "Self-Consistent Equations Including Exchange and Correlation Effects", Phys. Rev. **140**, A1133–A1138 (1965).

[Koopmans, 1934] T. Koopmans, "Über die Zuordnung von Wellenfunktionen und Eigenwerten zu den einzelnen Elektronen eines Atoms", Physica **1**, 104–113 (1934).

[Kouba et al., 2001] R. Kouba et al., "Phonons and Electron-Phonon Interaction by Linear-Response Theory within the LAPW Method", Phys. Rev. **B64**, 184306-1-9 (2001).

[Kubo, 1957] R. Kubo, "Statistical Mechanical Theory of Irreducible Processes, I", J. Phys. Soc. Japan **12**, 570–586 (1957).

[Labbe and Friedel, 1966a] J. Labbe and J. Friedel, "Effet de la température sur l'instabilité électronique et le changement de phase cristalline des composés du type V_3Si a basse température", J. Physique **27**, 303–308 (1966).

[Labbe and Friedel, 1966b] J. Labbe and J. Friedel, "Instabilité électronique et changement de phase cristalline des composés du V_3Si a basse température", J. Physique. **27**, 153–165 (1966).

[Labbe and Friedel, 1966c] J. Labbe and J. Friedel, "Stabilité des modes de distortion périodiques d'une chaine linéaire d'atoms de transition dans une structure cristalline du type V_3Si", J. Physique **27**, 708–716 (1966).

[Landau, 1941] L. D. Landau, "teorya sverkhtekuchesti geliya-II", Zh. Eksp. Teor. Fiz. (Russ.) **11**, 592–614 (1941).

[Landau, 1956] L. D. Landau, "The Theory of a Fermi Liquid", Sov. Phys. JETP **3**, 920–925 (1956).

[Landau and Lifshits, 1980a] L. D. Landau and E. M. Lifshits, *Statistical Physics, Part I* (Pergamon Press, London, 1980).

[Landau and Lifshits, 1980b] L. D. Landau and E. M. Lifshits, *Statistical Physics, Part II* (Pergamon Press, London, 1980).

[Laughlin, 1983] R. B. Laughlin, "Anomalous Quantum Hall Effect: An Incompressible Quantum Fluid with Fractionally Charged Excitations", Phys. Rev. Lett. **50**, 1395–1398 (1983).

[Lifshits, 1960] I. M. Lifshits, "Anomalies of Electron Characteristics of a Metal in the High Pressure Region", Soviet Phys. JETP **11**, 1130–1135 (1960).

[Lifshits et al., 1973] I. M. Lifshits, M. Y. Azbel, and I. M. Kaganov, *Electron Theory of Metals* (Consultants Bureau, New York, 1973), (German ed. Akademie-Verlag, Berlin, 1975).

[Lifshits et al., 1988] I. M. Lifshits, S. A. Gredeskul, and L. A. Pastur, *Introduction to the Theory of Disordered Systems* (Wiley, New York, 1988).

[Linde, 1985] A. Linde, "The Inflationary Universe", Rep. Prog. Phys. **47**, 925–986 (1985).

[Luttinger, 1960] J. M. Luttinger, "Fermi Surface and Some Simple Equilibrium Properties of Interacting Fermions", Phys. Rev. **119**, 1153–1163 (1960).

[Luttinger, 1961] J. M. Luttinger, "Theory of the de Haas-van Alphen Effect for a System of Interacting Fermions", Phys. Rev. **121**, 1251–1258 (1961).

[Madelung, 1978] O. Madelung, *Introduction to Solid State Theory* (Springer, Berlin, 1978).

[Maksimov, 1976] E. G. Maksimov, "A Self-Consistent Description of the Electron-Phonon System in Metals and the Problem of Lattice Stability", Soviet Phys. JETP **42**, 1138–1143 (1976).

[Maradudin, 1974] A. A. Maradudin, "Elements of the Theory of Lattice Dynamics", in *Dynamical Properties of Solids*, edited by G. K. Horton and A. A. Maradudin (North-Holland, Amsterdam, 1974), Vol. 1, Chap. 1.

[Martin and Schwinger, 1959] P. C. Martin and J. Schwinger, "Theory of Many-Particle Systems, I", Phys. Rev. **115**, 1342–1373 (1959).

[Migdal, 1958] A. B. Migdal, "Interaction Between Electrons and Lattice Vibrations in a Normal Metal", Sov. Phys. JETP **34**, 996–1001 (1958).

[Möbius and Richter, 1986] A. Möbius and M. Richter, "The Coulomb Gap in 1D Systems", J. Phys. C: Solid State Phys. **20**, 539–549 (1986).

[Mott and Davis, 1979] N. F. Mott and E. A. Davis, *Electron Processes in Non-Crystalline Materials* (Clarendon Press, Oxford, 1979).

[Nambu and Jona-Lasinio, 1961] Y. Nambu and G. Jona-Lasinio, "Dynamical Model of Elementary Particles Based on an Analogy with Superconductivity. I", Phys. Rev. **122**, 345–358 (1961).

[Norman and Freeman, 1986] M. R. Norman and A. J. Freeman, "Model Supercell Local-Density Calculations of the $3d$ Excitation Spectra in NiO", Phys. Rev. **B33**, 8896–8898 (1986).

[Ortuño et al., 2001] M. Ortuño, J. Talamantes, E. Cuevas, and A. Díaz-Sánchez, "Coulomb Interactions in Anderson Insulators", Phil. Mag. **81**, 1049–1064 (2001).

[Pitaevskii, 1976] L. P. Pitaevskii, "Weakly Bound Excitation States in a Crystal", Sov. Phys. JETP **43**, 382–388 (1976).

[Pollak and Pike, 1972] M. Pollak and G. E. Pike, "ac Conductivity of Glasses", Phys. Rev. Lett. **28**, 1449–1451 (1972).

[Schrieffer, 1964] J. R. Schrieffer, *Theory of Superconductivity* (Benjamin W. A., Inc., New York, 1964).

[Schwarz and Seiberg, 1999] J. H. Schwarz and N. Seiberg, "String Theory, Supersymmetry, Unification, and All that", Rev. Mod. Phys. **71**, S112–S120 (1999).

[Segal, 1947] I. E. Segal, "Postulates for General Quantum Mechanics", Annals of Mathematics **48**, 930–948 (1947).

[Sewell, 1986] G. L. Sewell, *Quantum Theory of Collective Phenomena, Monographs on the Physics and Chemistry of Materials* (Oxford University Press, Oxford, 1986).

[Shklovskii and Efros, 1984] B. I. Shklovskii and A. L. Efros, *Electronic Properties of Doped Semiconductors* (Springer-Verlag, Berlin, 1984).

Bibliography

[Smirnov and Tsvelik, 2003] F. A. Smirnov and A. M. Tsvelik, "Model with Propagating Spinons beyond One Dimension", Phys. Rev. **B68**, 144412–1–7 (2003).

[Steward, 1984] G. R. Steward, "Heavy-Fermion Systems", Rev. Mod. Phys. **56**, 755–787 (1984).

[Strange, 1998] P. Strange, *Relativistic Quantum Mechanics* (Cambridge University Press, Cambridge, U.K., 1998).

[Taylor, 1963] P. L. Taylor, "Theory of Kohn Anomalies in the Phonon Spectra of Metals", Phys. Rev. **131**, 1995–1999 (1963).

[Thirring, 1980] W. Thirring, *A Course in Mathematical Physics* (Springer-Verlag, New York, 1980), Vol. 4.

[Valatin, 1958] J. G. Valatin, "Comments on the Theory of Superconductivity", Nouvo Cimento **7**, 843–857 (1958).

[von Neumann, 1955] J. von Neumann, *Mathematical Foundation of Quantum Mechanics* (Princeton University Press, Princeton NJ, 1955).

Index

$SO(3)$, 19
$SU(2)$, 19

adiabatic approximation, 155
anomalous mean values, 140
anti-commutator, 36

band model, 118
Bloch electrons, 118
Bogolubov's quasi-means, 143
bootstrap principle, 135
Born-van Kármán boundary conditions, 24
boson gas, interaction-free, 99
Bragg scattering, 119
Breit interaction, 150
Brillouin zone, 119

canonical transformation, 96, 104–109
charge operator, 56
charge space, 138
chemical potential, 72
closed-shell Hartree-Fock method, 30
coherent state, 38
coherent state representation, 40, 46
collision integral, 116
commutator, 35
complete set, 14, 34
    of N-particle states, 28
    of spin-orbitals, 16
complex conjugate, 15
conjugate Grassmann numbers, 44
convex, 59
Cooper pair, 104
Coulomb gap, 130
Coulomb glass, 129
creation and annihilation operators, 37
    bosonic, 35
    coherent state representation, 40, 46
    fermionic, 36
cyclic vector, 36

densities of macrosystems, 56

density functional, 157
density matrix, 60
    canonical, 65
density of states, 71, 81, 85–87
density operator, 47, 56
density parameter, 123
derivative of a quantum field, 49
Dirac-Coulomb Hamiltonian, 150
Dirac-Kohn-Sham equation, 158
discrete translational symmetry, 119
dispersion law, 114
dispersion of an observable, 58
dispersion relation, 91, 114
double-well potential, 131
Drude model, 118
Dyson equation, 122, 126
Dyson's formula, 80

Ehrenfest relation, 120
eigenstate, 13
eigenvalue, 14
electron Fermi surface, 121
electron transport, 121
electron-phonon interaction, 127
empirical pseudopotentials, 119
ensemble, 64
exchange and correlation hole, 123
exchange energy, 31
exchange potential operator, 32
exchange term, 26
expectation value, 13
extremal point, 61

f-levels, 131
Fermi radius, 28
Fermi surface, 119
fermion gas, interaction-free, 85, 95
    homogeneous, 27
field operators, 47
Fock operator, 32
Fock space, 34, 135

Index 171

generalized, 44

gauge field, 137
Gelfand-Naimark-Segal construction, 61, 137
global gauge transformation, 138
global operator, 56
global stability, 66
gluon exchange, 151
GNS construction, 61
Goldstone mode, 135
Green's function, 77
    causal, 77
    Fourier transformed, 78
    matrix, 81
    Matsubara, 92
    of non-interacting bosons, 79
    of non-interacting fermions, 78
    ordinary, 77
    retarded and advanced, 80, 90
    spectral representation, 79

Hamiltonian, 14
    in Heisenberg representation, 29
    as an operator form, 73
    field quantized, 48
    grand canonical, 72
    in Schrödinger representation, 18
    in momentum representation, 25
    in occupation number representation, 37
harmonic lattice, 125
Hartree energy, 31
Hartree potential, 32
Hartree-Fock energy, 31, 33
Hartree-Fock equations, 32
Hartree-Fock orbitals, 32
heavy gauge bosons, 151
Heisenberg picture, 14, 74
Herglotz function, 80
Hermitian operator, 13
Higgs mode, 135
Hilbert space
    N-particle, fermionic, 17
Hilbert space, 13

N-particle, bosonic, 17
hole Fermi surface, 121
hole transport, 121

interaction picture, 74
inverse mass tensor, 121
irreducible representation space, 36

kinetic equation, 116
KMS condition, 65
Kohn-Sham equation, 158
Kohn-Sham orbital energies, 159
Kohn-Sham orbitals, 157
Koopmans' theorem, 33
Kubo formula, 89
Kubo-Martin-Schwinger condition, 65

Landau's quasi-particle interaction function, 116
left eigenstate, 39
Lehmann representation, 79, 91
linear response, 89
linear, positive and normalized function, 59
local approximation for the self-energy, 124
local density approximation, 158
local stability, 66
locally finite, 58
Luttinger theorem, 125

mean field, 32
measure $d\mu_c$, 46
metal-insulator transition, 129
metastable state, 67
mixed state, 60
model Hamiltonian, 110
molecular field, 32
multi-particle excitations, 115

normal coordinates, 125
normal order, 37
normalization volume, 24
nucleons, 151

observable, 13, 49, 50
    macroscopic, 58
    microscopic, 58

occupation number, 34
off-diagonal long-range order, 144
open-shell Hartree-Fock method, 30
operator algebra, 36, 59
orbital relaxation energy, 33

particle density, 22, 28
particle number operator, 35, 72
partition function, 65
Pauli matrices, 15, 18
periodic boundary conditions, 24
permanent, 17
pion exchange, 151
polarization operator, 126
propagator, 78
pure state, 60

quantum crystals, 126
quantum fields, 47, 49
quantum fluctuation, 13
quarks, 151
quasi-momentum, 119
quasi-particle renormalization parameters, 129
quasi-particle conception, 115
quasi-particle density of states, 86, 122, 129
quasi-particle energy, 115
quasi-particle interaction, 115
quasi-particle spectral amplitude, 123
quasi-particle spectral function, 122
quasi-particle spectral weight, 122
quasi-particle vacuum, 114

reciprocal lattice vector, 119
reducible representation space, 100
relativity principle in the charge space, 137
renormalization constant, 82

Schrödinger picture, 14, 73
self-averaging, 129
self-energy operator, 122
self-interaction, 31
sesquilinear form, 138

short-range correlations, 63
single-particle excitations, 114
Slater determinant, 16
Sommerfeld's theory of metals, 118
spectral density of states, 87
spectral representation, 79
spin density matrix, 22–23
spin density matrix operator, 48
spin flip amplitude, 23
spin flip operator, 48
spin operator, 15
spin-orbital, 16
spinor, 15
state, 62
statistical ensemble, 67
statistical operator, 65
statistics sign factor, 46
super-analysis, 42
super-symmetry, 42, 47
superselection rules, 54
superselection sectors, 54
superstrings, 152
susceptibility, 89

thermodynamic limit, 56
thermodynamic stability condition, 66
torus, 24

vacuum polarization, 115
vacuum state, 36
valence fluctuation, 132
variational principle by Hohenberg and Kohn, 157
vector spin, 22–23
vector spin density, 22–23
vector spin density operator, 48
vibrational motion, 125

wave pocket, 42, 98, 119
wavenumber space, 24
weakly non-ideal gas, 70

zero-vibrations, 126

Aus dem Verlagsprogramm:

Franeck, H. (Freiberg / Dresden):
EAGLE-STARTHILFE Technische Mechanik.
Ein Leitfaden für Studienanfänger des Ingenieurwesens.
EAGLE 015: www.eagle-leipzig.de/015-franeck.htm ▶ ISBN 3-937219-15-3

Klingenberg, W. P. A. (Bonn):
Klassische Differentialgeometrie.
Eine Einführung in die Riemannsche Geometrie.
EAGLE 016: www.eagle-leipzig.de/016-klingenberg.htm ▶ ISBN 3-937219-16-1

Luderer, B. (Chemnitz):
EAGLE-GUIDE Basiswissen der Algebra.
Reihe: EAGLE-GUIDE / Mathematik im Studium (Hrsg.: B. Luderer)
EAGLE 017: www.eagle-leipzig.de/017-luderer.htm ▶ ISBN 3-937219-17-X

Fröhner, M. / Windisch, G. (beide Cottbus):
EAGLE-GUIDE Elementare Fourier-Reihen.
Reihe: EAGLE-GUIDE / Mathematik im Studium (Hrsg.: B. Luderer)
EAGLE 018: www.eagle-leipzig.de/018-froehner.htm ▶ ISBN 3-937219-18-8

Sprößig, W. (Freiberg) / Fichtner, A. (München):
EAGLE-GUIDE Vektoranalysis.
Reihe: EAGLE-GUIDE / Mathematik im Studium (Hrsg.: B. Luderer)
EAGLE 019: www.eagle-leipzig.de/019-sproessig.htm ▶ ISBN 3-937219-19-6

Resch, J. (Dresden):
EAGLE-GUIDE Finanzmathematik.
Reihe: EAGLE-GUIDE / Mathematik im Studium (Hrsg.: B. Luderer)
EAGLE 020: www.eagle-leipzig.de/020-resch.htm ▶ ISBN 3-937219-20-X

Thierfelder, J. (Ilmenau):
EAGLE-GUIDE Nichtlineare Optimierung.
Reihe: EAGLE-GUIDE / Mathematik im Studium (Hrsg.: B. Luderer)
EAGLE 021: www.eagle-leipzig.de/021-thierfelder.htm ▶ ISBN 3-937219-21-8

Günther, H. (Bielefeld):
EAGLE-GUIDE Raum und Zeit – Relativität.
EAGLE 022: www.eagle-leipzig.de/022-guenther.htm ▶ ISBN 3-937219-22-6

Edition am Gutenbergplatz Leipzig: www.eagle-leipzig.de

Aus dem Verlagsprogramm:

Bandemer, H. (Freiberg / Halle a. d. Saale):
Mathematik und Ungewißheit.
Drei Essais zu Problemen der Anwendung.
EAGLE 023: www.eagle-leipzig.de/023-bandemer.htm ▶ ISBN 3-937219-23-4

Eschrig, H. (Dresden):
The Particle World of Condensed Matter.
An Introduction to the Notion of Quasi-Particle.
EAGLE 024: www.eagle-leipzig.de/024-eschrig.htm ▶ ISBN 3-937219-24-2

Ortner, E. (Darmstadt):
Sprachbasierte Informatik.
Wie man mit Wörtern die Cyber-Welt bewegt.
EAGLE 025: www.eagle-leipzig.de/025-ortner.htm ▶ ISBN 3-937219-25-0

Britzelmaier, B. (Pforzheim):
EAGLE-STARTHILFE Finanzierung.
EAGLE 026: www.eagle-leipzig.de/026-britzelmaier.htm ▶ ISBN 3-937219-26-9

Siehe auch:

Eschrig, H. (Dresden):
The Fundamentals of Density Functional Theory.
2nd Edition.
EAGLE 004: www.eagle-leipzig.de/004-eschrig.htm ▶ ISBN 3-937219-04-8

Edition am Gutenbergplatz Leipzig: www.eagle-leipzig.de